COULTER LIBRARY ONON
HD9539.U7M67 1982
Moss, Norman. The politi
3 0418 000

D0150958

HD
9539
.U7
M67
1982
Moss, Norman
The politics of
uranium
DATE DUE
MAY 1 3 1983
COMMUNITY
0165 01 714557 01 0 (IC=0)
MOSS, NORMAN
POLITICS OF URANIUM
(4) 1982 . HD 9539 .U7 M67 1982
The Sidney B. Coulter Library
Onondaga Community College
Rte. 173, Onondaga Hill
Syracuse, New York 13215

Norman Moss is a writer and journalist. He has worked for both British and American news media, has been a foreign correspondent for an American radio news service, and his articles have appeared in many leading magazines. This is his fourth book; his first was a widely-praised political, scientific and moral history of the hydrogen bomb, *Men Who Play God*. He is married with two sons.

The Politics of Uranium

Norman Moss

THE POLITICS OF URANIUM

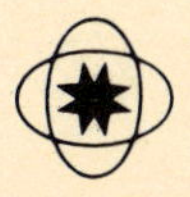

UNIVERSE BOOKS
New York

Published in the United States of America in 1982
by Universe Books
381 Park Avenue South, New York, N.Y. 10016

82 83 84 85 86/10 9 8 7 6 5 4 3 2 1

Printed in Great Britain

Library of Congress Cataloging in Publication Data

Moss, Norman, 1928–
The politics of uranium.

Bibliography: p. 226
Includes index.
1. Uranium industry – Political aspects.
2. Atomic energy industries – Political aspects.
3. Nuclear nonproliferation. 4. World politics.
I. Title.
HD9539.U7M67 333.8'5 81-21888
ISBN 0-87663-390-4 AACR2

Contents

Acknowledgments

The people who allowed themselves to be interviewed and gave me their time in the course of my researches for this book are far too numerous to be named individually, and besides, many of them would not want to be named because of the positions they hold or held recently. However, it would be ungracious not to mention a few people who have not been participants in the events recounted, but who have been generous in sharing with me their hard-earned knowledge and their time.

Among these are Terence Price, the Director of the Uranium Institute, and several of his associates, as well as the Institute's library staff; Walter Patterson, the physicist with Friends of the Earth, a critic of nuclear power who has always been ready to answer my questions patiently and sometimes repeatedly without requiring agreement with his viewpoint; Ian Smart, who has moved from diplomacy to academic life to consultancy, and has an unsurpassed knowledge of the nuclear power field; David Potter, the physicist, lately of the Imperial College of Science and Technology in London, who elucidated for me the nature of the universe and related topics on a number of evenings in a number of pubs; Professor Jack McFadean of the Department of Geology at Williams College, Massachusetts; and Professor Ashok Kapur of the Political Science Department at Waterloo University, Ontario. Naturally, none of these can be held responsible for anything in this book in their areas of expertise.

My thanks are due also to the Denison Mining Corporation for taking me down one of its uranium mines. Finally, I owe a debt of gratitude to my publisher, Piers Burnett, for his belief in this project, his encouragement and his patience.

Introduction

Uranium is a uniquely strategic commodity, being both a source of energy in a world acutely sensitive to energy needs, and the material for the most destructive explosive ever invented. Nuclear power is uranium power. It is not produced by a nuclear reactor, but in one. The uranium produces it all by itself, fissioning spontaneously and giving off heat and energy. It does this even dispersed in rock, in the bowels of the earth, contributing to our planet's inner heat and its turbulence below the surface. A nuclear reactor simply draws off the heat once the uranium is assembled and placed inside it. It is a device for exploiting the special properties of uranium, as a camera is a device for exploiting the special properties of film. The crucial questions concerning nuclear power are about the distribution of uranium and its by-products.

Decisions must be taken in the next few years that are of enormous import for all our futures, concerning our reliance on nuclear energy, the breeder reactor, the proliferation of plutonium, and measures to check the probable acquisition by more countries of atomic weapons. The consequences of these decisions will be with us for centuries, at least. If they are wrong ones, then the sins of the fathers will indeed be visited upon the children even unto the third and fourth generations. These could be sins of omission as well as commission, for deciding to abstain from one kind of nuclear power is also a decision with far-reaching consequences, for better or worse; it could produce results opposite to those intended.

Some of the decisions will be technical, but they will be based on political, economic and social analyses and choices. These will concern questions of values, such as the relative importance we attach to material prosperity and international safety, and our role as the custodians of this planet as well as its tenants.

This book deals with the part that uranium plays in the life of the world today, questions arising from its use, principally those involving the relationship between nuclear power and nuclear weapons, and the way these questions have been answered so far. Telling this story, several interesting points emerge:

One is the way that what appears to be a technical issue can very quickly become political as well. The simplest example of this is the question of how much uranium we can find in the ground. The answer to this question can be a factor – in fact, always is a factor – in governmental decisions on the breeder reactor, which can extend the useful life of uranium; and plutonium, which can sometimes serve as a substitute; and measures to check the spread of nuclear weapons. A predilection for a particular policy in these areas may influence the estimates of the availability of uranium. Or, to take another example, the choice between two types of nuclear power plant, ostensibly based on efficiency and cost, may in fact depend on how willing a country is to depend on another for its fuel supply, and whether it wants to keep open the option to build an atomic bomb.

Another is the shortsightedness of so many people involved in nuclear power in the past about the consequences of what they were doing. The nuclear power programmes in some Third World countries that are causing anxieties among the big powers today were set in motion enthusiastically by some of those same powers. There is very little in the early literature of nuclear power about the problem of nuclear waste, or the possibility of an accidental release of radioactivity, subjects which are causing much concern today.

Another is the extent to which changes have been brought about by new perceptions and attitudes rather than new technology. The widespread controversy about a breeder reactor and a 'plutonium economy' might convey the impression that the breeder reactor is a new development. Actually, the idea of the breeder is almost as old as the idea of nuclear power, and a model breeder was operating more than twenty-five years ago. In those days, it was seen as the reactor of the near future, and hardly anyone worried about the prospect. It is the controversy and concern that are new, not the technology or the knowledge. The same applies to the current controversies over the reprocessing of used fuel. There are no problems confronting us that have been brought about by new technical developments in nuclear fission since 1945, though there may be some soon.

Still another is the international interdependence of nuclear power. Few countries are blessed with the natural resources, the technological capability and the money that are required to develop a nuclear power programme unaided. Most rely on some other country to supply one or more of these. Hence, there are many points at which leverage can be exerted on the way that nuclear power is used. When Australians debate the morality of mining uranium, Western Europeans watch anxiously, as potential purchasers. Japan must get permission from the United States before sending its used fuel to Britain or France for reprocessing, and permission must be given by an Executive waiver under an act of Congress. Germany asks Canada whether it may send the Canadian uranium it has bought for its nuclear power plants to Russia first, to be enriched there. This sometimes makes the politics of uranium complicated.

Finally there is what I will call an incongruity of scale. Following a question involving nuclear power point by point, one suddenly steps off an edge, as it were, and plunges into a question on a totally different scale of importance, or of time. Thus, consideration of the benefits and cost of one kind of nuclear fuel cycle as against another leads abruptly to a consideration of whether one kind will make a nuclear holocaust more likely. Discussion of energy needs in twenty-five years' time suddenly becomes a discussion of the storage of radioactive materials over 100,000 years.

The aim is to illuminate present-day problems and some future ones. However, much of the present situation is best seen through the events and sometimes the mistakes that produced it. So some of the history of uranium is told here, beginning at the beginning.

Chapter 1

WHAT GOD HATH JOINED . . . ?

The universe began between 10 and 15 billion years ago, according to the astronomers' canonical history, with an explosion of energy much of which turned into matter almost immediately. Not only energy and matter began then, but space and time also, something we cannot conceive in terms of our own experience of space and time. To bring this idea out of the realm of the mystical and into that of more recognizable scientific statements, cosmologists sometimes express it more cautiously, and say that at that moment there was a singularity, a once-and-for-all event, and that our knowledge of space and time and the universe can reach back only to that, and no further.

The matter that was created out of the primary energy took the form of three fundamental particles that make up most of matter today: the proton, the electron and the neutron. The electron carries a negative electrical charge; the proton carries a positive charge and is 1,840 times as big as an electron. The neutron is as big as the two combined and carries no electrical charge.

The protons and neutrons joined together very quickly. For the next 100,000 years or so, these and the electrons were loose in what physicists call a plasma, a gas so hot that it consists not of elements, but only of sub-atomic particles, each unconnected with the others, racing about at great speed. Then, after the first thousand centuries, as the universe (and space) continued to surge outwards to cool, this primeval chaos acquired form: the particles drew together to make up atoms. Protons drew electrons towards them, so that groups of electrons came to spin in orbit around the same number of protons, according to complex patterns.

The atoms formed elements, each kind of atom forming a different element. The characteristics of each element were determined by the number of protons in the nucleus of its atom. This is the material universe that we know today.

There is a special force, powerful over very short distances, sometimes called the binding force, that pulled these protons and neutrons together and keeps them together. It must be powerful since the protons all have the same electrical charge, which means that electrical forces tend to push them apart, and this force must be overcome. When a large number of protons are gathered together, their combined electrical repulsion overcomes the binding force after a while, so that the atom throws off a particle and becomes a different kind of atom. We call this process radioactivity. An element that undergoes this process does so at a regular rate, that is, a fixed proportion of the atoms change in a certain period of time. We measure this rate in half-lives, a half-life being the period in which half the number of that particular kind of radioactive atom will have disintegrated into something else.

Our solar system began between four and five billion years ago, as a cloud of swirling hot gas that somehow broke away from the rest to do its own cosmic dance and form its own patterns, a part of the larger patterns of the Milky Way galaxy. This cloud of gas coalesced, and as it did so, the heat was so intense that the nuclei of atoms were crushed together, and some smaller atoms were formed into larger ones, a process that takes place wherever stars are created. The main part of the gas cloud became the sun, and separate but smaller clouds broke away to become the nine planets that revolve around it.

The heaviest element in this cloud is also the heaviest that exists naturally anywhere, so far as we know. It has 92 protons, and the same number of electrons, and anywhere from 141 to 146 neutrons. It is slightly radioactive and so it is disintegrating, but at a slow rate. Its half-life, as it happens, is 4½ billion years, which is just about the age of the solar system, so that half the amount that was present when the solar system began remains today. This is uranium.

* * *

There is quite a lot of uranium on the Earth. It is, for instance, a thousand times as plentiful as gold, and nearly as plentiful as tin, nickel and zinc. One 250,000th of the Earth's crust is uranium. However, the arrangement of the electrons in the outer orbit is such that there are gaps into which electrons of other atoms can fit snugly. Because of this, it combines easily with other elements, and forms many compounds. It is distributed in minute quantities nearly everywhere on land and sea – there is some in all sea water.

The Earth was molten at first, but it cooled over the eons, and much of it solidified. But its form was not then fixed and immutable. To us, its passengers, spaceship Earth seems a stable craft, with the ground solid beneath our feet. But a speeded-up motion picture of its history would show it as a sphere undergoing convulsions, with the explosive gases at its centre pushing the surface this way and that so that the surface heaves and plunges, with the continents moving about and breaking up, the high becoming low and the low high, the mountain a valley and the ocean a desert.

Two billion years ago, when the contours of the Earth's surface bore little resemblance to those of today, underground streams picked up specks of uranium, because uranium combines easily with water, and carried them to the surface, mixing them along the way with specks of quartz, zircon and granite. These formed themselves into gritty rocks, with jagged surfaces, dark grey in colour but glittering with pyrite. Geologists call these quartz conglomerates. Over the years since then, while the two great land masses broke up into the present continents, and new mountains bulged out of the earth and oceans changed their shape, some of these quartz conglomerates remained on the surface and remained unchanged, apart from the slight, steady depletion of the uranium content by radioactive decay. They remained unchanged and even undisturbed while plant life flourished and animal life appeared and then Man, and then civilization, until, by a lakeside in Ontario and in the Australian desert, a few prospectors picked up some of them and examined them closely and excitedly, and knew that they had found what they were looking for.

More than a billion years after these quartz conglomerates were formed, another kind of uranium deposit was created by a quite different process. Blobs of minerals were thrust upwards through the rocks in the Earth's crust by subterranean pressures, and they drew into themselves several metals that combine easily with one another: copper, manganese, zircon, sillicon and uranium. These are usually smaller than other deposits, but the concentration of metals, including uranium, is higher. These blobs reached the surface in central Africa, in what is now Zaire, and the Canadian North, and Czechoslovakia, and France. Though sometimes streaked with bright colours that denote the presence of uranium and other metals, the rock is basically black and so it is called pitchblende.

Later still, uranium deposits were created by another kind of movement entirely: the draining away of a huge inland sea from what is now the rocky and mostly arid western United States. As the waters drained away, they carried along with them specks of uranium and some other minerals out of the rocks. By now, there had been life on Earth for more than a billion years, and long-decayed organic matter was now buried under layers of rocks. This worked a chemical process that drew the particles of uranium out of the water, along with particles of some other minerals, and separated them, and left them to adhere to the sides of rocks as the waters drained away. These particles formed a mineral that geologists today call carnotite. It stained the yellow sandstone rock with patches of rust-red, blue and green, and would one day be made into war paint by the Apache and Comanche Indians who would inhabit this land. Patches of carnotite are dotted around the area known to geographers as the Colorado Plateau, covering parts of Colorado, Utah, Wyoming and New Mexico.

The ability of uranium to combine with many different minerals, and to move with any flow of earth or water, was to be a headache for geologists when it became a sought-after metal. Thus, L.C. Jacobsen, in a paper on uranium supply published in 1979, wrote:

> For most minerals, nearly all significant deposits fit into one of a small number of genetic models. Application of this geologic knowledge is usually sufficient to narrow the area of exploration for any given mineral to specific geologic provinces. . . . Uranium, in contrast to such minerals as copper, lead, nickel, and so on, has a multiplicity of models. In theory, it can occur almost anywhere. It is found in young rocks, old rocks, sedimentary, metamorphic and igneous rocks, mountains and plains. The great variety of uranium occurrences enormously complicates exploration; few areas can be rejected out of hand.*

Uranium was discovered as a separate element in 1789, by the pioneer German chemist Martin Klaproth, who also discovered titanium, zircon and cerium. It was the eighteenth element to be

*US Uranium Price and Supply. In *Materials and Society*, 1979.

identified out of the ninety-two that we now know exist in their natural state. Klaproth named the new element after the planet Uranus, which had been discovered eight years earlier. He could not know that when all the other elements were identified, uranium would turn out to be the heaviest.

Klaproth found his uranium in pitchblende in the metal mines at Joachimsthal, in the Erz Giberg Mountains in Bohemia, in what is now Czechoslovakia. The mines were 250 years old even then. The silver mined there was so pure that it was at one time the standard coinage for much of central Europe. A coin made from its silver was known as a Joachimsthaler, or, later, a thaler, pronounced with a short 'a'; the word 'dollar' derives from this. Uranium metal is yellow, and it has a strong colour, so its powder was found to be useful as a colouring agent in the production of glassware and pottery. In the late nineteenth century, a few tons of it were extracted from the Joachimsthal mines each year and used for this purpose. It had no other use.

In the early years of this century, Marie Curie discovered radioactivity in radium. Lesser amounts of radioactivity were found to exist in other elements as well, including uranium. Radium is created by the decay of uranium, so where uranium is found, there is always some radium as well. When Mme Curie wanted to isolate some radium, she had sackfuls of pitchblende from Joachimsthal, from which uranium had already been extracted, sent to her laboratory in Paris, and she and her husband combed through this. Later, Joachimsthal became a major source of uranium for the Soviet Union.

It was found that radium had powerful medicinal properties even in very small quantities. It was sought after, and because of its rarity, it became by far the most expensive metal in the world. (Radium is no longer much used in medicine; today radioactive isotopes created in laboratories, or else powerful and precise x-ray machines, can achieve similar effects.) Now uranium was sought, simply because wherever it was found, there radium would be found also. The quantities of radium would be minute, usually one part radium to three million parts uranium, but it could still be worth while. Small quantities of radium were mined in Australia, at Mount Painter and the optimistically named Radium Hill. Radium was found and mined in the carnotite in Colorado, along with vanadium, a metal very useful for hardening steel.

In April 1915, a surveyor in the Belgian Congo was looking for ores in the concession area in Katanga Province given to the Union Minière de Haute Katanga. He told his African assistant to dig in a hillside near the village of Shinkalobwe, and found pitchblende, streaked with colours that indicated to him the presence of a number of valuable metals, including uranium. He reported his find immediately in a telegram to his head office, specifying the presence of what he called 'a radioactive material'. In wartime, no capital could be found to start a big new mining operation. Mining started in 1921, in secret, and ore was shipped to a refinery in Oolen, Belgium, where the radium was extracted, still in secret. In January 1923, the Union Minière broke cover with a gift of radium to Belgian hospitals. Soon, Shinkalobwe was producing radium that was selling at $100,000 a gramme.

Expensive though this was, it undercut the price of Colorado radium, and the Colorado radium mines closed down. Output was measured in grammes: in 1929, a record high was reached with the production of 60 grammes at Shinkalobwe. There was not much use for the uranium left in the ore after the radium had been removed, but each year about 80 tons was extracted and sold to glass and pottery firms, where it was used to add colouring, and a small amount was sent to laboratories and geological museums.

The Belgian Congo's near-monopoly lasted about ten years. It was broken by a discovery in the Canadian Arctic. Gilbert LaBine was an adventurous prospector and miner who already had his own company operating successfully in other parts of Canada, Eldorado Gold Mines Ltd. In 1929, following up a lead suggested in an explorer's report thirty years earlier, he had a seaplane take him up to the Great Bear Lake, a huge lake in the Northwest Territory bisected by the Arctic Circle, to look in the cliffs for pitchblende. He found it, with traces of a high uranium content, which indicated the presence of usable amounts of radium. His company started mining, at a site which he christened Port Radium. He built a refinery to handle the ore 2,000 miles away at Port Hope, on the shore of Lake Ontario, and sent the ore there by road and rail. The Port Hope refinery produced its first 3 grammes of radium in 1932. By 1938, it was producing 85 grammes in a single year, and had forced the world price down to half of what had been the

Shinkalobwe price, which was bad news for Union Minière. Like Shinkalobwe, the Port Hope refinery sold a small amount of uranium to the pottery and glass industries, but most of it remained in the tailings, which is what miners call the residue of the ore.

* * *

In the late 1930s, two quite separate activities, in two very different areas of human endeavour came to a head, in a way that was to give uranium a new importance. One was the series of discoveries in physics that led to an understanding of the phenomenon of nuclear fission. The other was the current in international relations that led to the Second World War.

Among physicists, the discovery of radioactivity opened the way to the exploration and manipulation of the nucleus of the atom. In Rome in the 1930s Enrico Fermi bombarded many different kinds of atoms with neutrons, knocking away one or two particles from an atom to create a new isotope, a new kind of atom. He was puzzled about what happened when he bombarded uranium with neutrons, and no one else could understand it either. This kind of experiment is conducted with minute quantities of matter, and the immediate effects take place in a microcosm so small as to be far beyond the reach of human observation. As with all atomic physics, what happens inside an atom must be worked out, or guessed at, from analysis of the observed effects.

In 1938, two German chemists, Otta Hahn and Fritz Strassmann, analysed the results of Fermi's bombardment of uranium, and announced that they had found traces of barium. This was baffling, because the atomic weight of barium is a little more than half that of uranium, and it could hardly be produced by knocking one or two particles out of uranium atoms. Hahn and Strassmann published their findings in a scientific journal, leaving it to someone else to find an explanation. It seemed inexplicable, and some physicists suggested that their chemical analysis was mistaken, and it was not really barium they had found.

Hahn also wrote an account of his findings in a letter to a former colleague with whom he was still in touch, Lise Meitner, an Austrian physicist who had been forced to leave Nazi Germany because she was Jewish, and was now at the Nobel Institute in Stockholm. The letter arrived at Christmas-time 1938, when Dr

Meitner was entertaining her nephew, Otto Frisch. Frisch was also a nuclear physicist who had left Germany, and was then at the Niels Bohr Institute in Copenhagen. The two pondered the matter together during a long walk through the snow-covered countryside. Then Dr Meitner had an idea: she suggested that some very heavy atoms might be unstable in an unusual way, so that one could be split into two roughly equal parts by the injection of a neutron, and this could be how barium had resulted from the neutron bombardment of uranium. They both sat down on a log and tested this possibility mathematically on the spot by working out the calculations (for the process is much more complicated, and more profound in its implications, than is indicated by this mechanistic summary).

Frisch returned to Copenhagen after Christmas and told Niels Bohr, then the grand old man of atomic physics, about their idea. Bohr struck his forehead and said, 'What idiots we've all been! This is just the way it must be.' He urged Frisch to write a paper on this at once, which Frisch did, collaborating with his aunt on the telephone.

He asked an American biologist working at the Bohr Institute what biologists called it when a living cell divided in two, and he was told that the term was 'fission'. So Frisch coined the term 'atomic fission', and used it in the paper that he and Lise Meitner sent to the British scientific journal *Nature*. As soon as this paper appeared, atomic physicists in other centres repeated Fermi's experiments with uranium and confirmed the effects. In America, they were doing this sooner, because Bohr went on a visit to America very soon after his conversation with Frisch, and in his excitement he told American scientists about the idea.

Curiously, Frisch and Lise Meitner missed a vital point. This was that if a uranium nucleus is split by the impact of a neutron, then other neutrons from this atom are sent shooting off, and these may in turn split other uranium atoms. The point was spotted by some others. The Hungarian-born physicist Leo Szilard, then at Columbia University in New York, saw this, and some other implications as well. In February 1939, he wrote to Fréderic Joliot-Curie, Marie Curie's son-in-law, about his reaction to these new discoveries.

> A few of us got interested in the question of whether new neutrons are liberated in the disintegration of uranium. Obviously, if more than one neutron was liberated, a sort of chain reaction might be possible. In certain circumstances, this might then lead to the construction of bombs, which would be extremely dangerous in general and particularly in the hands of certain governments.

When the nucleus of a uranium atom is split, the binding energy is released. The amount of energy is miniscule, but if a chain reaction takes place so that a lot of atoms undergo fission, then the cumulative amount of energy released could be enormous.

At the beginning of 1939, it was natural for Szilard to see this in terms of an explosive and of danger. Germany, the biggest country in central Europe, had succumbed to Nazism, annexed Austria, and was now swallowing Czechoslovakia brutally in two bites. Every major power was rearming. War seemed to be looming and it was forecast that aerial bombardment on an unprecedented scale would accompany it.

Joliot-Curie was not slow to take up the point about a chain reaction. In March, he and Lew Kowarski and the Austrian-born Hans Halban demonstrated in their laboratory in Paris that extra neutrons were indeed liberated in the splitting of a uranium atom. In May, he took out three patents in the name of the French national laboratory. Two were for the production of power by atomic fission; one was for the creation of a fission explosion.

He set about making plans for a project that might culminate in a test explosion in the Algerian Sahara, underestimating enormously the difficulties involved. For this he needed uranium. Joachimsthal was now in German hands, so he turned to the Belgian Congo mines, that seemed to contain quantities of uranium, the hitherto unwanted stuff that accompanies radium. He approached the Union Minière de Haute Katanga on behalf of his laboratory, and proposed that they collaborate on this experiment. They drew up a contract for co-operation. It was initiated by Joliot-Curie on behalf of his laboratory, and by the Chairman of Union Minière, Edgar Sengier. The contract was never completed because war broke out four months later, and with Belgium neutral, Union Minière did not want to be tied to a French military project. The agreement had called for collaboration on the project, and said that if it were successful, the two parties would jointly exploit the results, civil or military. Union Minière was to contribute funds, and also 50 tons of uranium.

For the first time, someone sought uranium in quantity for its own properties, and not merely because of its proximity to radium. During the next few years, the location and control of uranium were to be matters of speculation, anxiety and controversy.

* * *

The coming of war stimulated interest in uranium fission explosion, even though the possibility seemed remote and there were competing demands on resources. In Paris, Joliot-Curie, Halban and Kowarski worked out some of the requirements for a chain reaction. They even acquired some heavy water, water which contains an isotope of heavy hydrogen instead of ordinary hydrogen, which they would need for a preliminary experiment. Later, Halban and Kowarski were to work on the atom bomb project in Canada.

In the United States, Szilard thought that the possibility of a fission explosive was obvious, and like many others, he feared that Germany might develop it before anyone else. He and Eugene Wigner, a fellow-Hungarian who shared his anxieties, went to see Albert Einstein, to ask him to lend his uniquely distinguished name to a letter to President Roosevelt, suggesting that the possibility be investigated. Roosevelt set up a Committee on Uranium to look into this, headed by Lyman J. Briggs, the Director of the National Bureau of Standards.

Otto Frisch went to England in 1939, to teach at Birmingham University. There, he pondered the question of a uranium explosion with Rudolf Peierls, a German, another refugee from Nazism. Frisch said at first that it would be a practical impossibility. This was because most uranium does not fission when hit by a neutron, only an isotope, which makes up less than 1 per cent of it. An isotope is a variation of an element which is chemically indistinguishable from the rest, with the same number of protons and electrons, but with slightly more or slightly fewer neutrons in its nucleus. The atomic weight of an element is measured by the number of protons and neutrons, so this varies in an isotope. Most uranium is u-238, but the part that fissions is an isotope that has the number 235, and makes up only 0.7 per cent of all uranium. Frisch explained that an explosive chain reaction could not take place because most of the neutrons that would be flying about would be stopped by the ordinary uranium atoms before they reached the rare atom of U-235.

But what, they then asked, if it were possible to separate the uranium 235 from the rest? Would a chain reaction be possible then? And how much uranium 235 would be required? They worked on this, and came to the startling conclusion that uranium 235 could indeed be separated, and 5 kilogrammes of it could create an explosion equivalent to several thousand tons of t-n-t. They sent a memorandum reporting these conclusions to the War Cabinet's scientific office in March 1940. The following month, the Government set up a committee to examine the prospects. In his memoirs, published posthumously in 1979, Szilard gives credit to the British for being the first to recognize that it would be possible to separate a sufficient quantity of uranium 235 to make a bomb, and for alerting the American authorities to this crucial fact.

In the wartime atomic bomb project, which became a joint US-British-Canadian project, uranium was the one essential material.

A key figure was Edgard Sengier, the Chairman of the Congo company Union Minière. Sengier always saw himself as something of a statesman as well as a businessman, with a statesman's responsibilities. He said a number of times, in conversation and not on oratorical occasions, that he devoted himself more to the welfare of the Africans in Katanga Province than to making a profit for his company. He took naturally to negotiating the uranium contract with Joliot-Curie, representing an agency of the French Government, on a government project.

He was no scientist, but he knew from these negotiations that uranium might have some military significance. When the Germans conquered Belgium in 1940, he was afraid that they might somehow gain control of the Belgian Congo also. He took it upon himself to remove the 1,200 tons of uranium ore that was piled up at the Shinkalobwe mine, and ship it to America. It was in steel drums marked 'Uranium Ore. Product of the Belgian Congo'; these were unloaded at a dock on Staten Island, in New York harbour, and they remained in a warehouse there. Sengier also removed himself to America, and established himself in Union Minière's New York office.

The Einstein letter to Roosevelt mentioned that uranium was to be found in the Belgian Congo and Canada. When the Manhattan Project, as the atom bomb programme came to be called, got under way in 1942, Colonel Kenneth Nichols, the deputy to the director of the project, was sent to the Union Minière office in New York to ask to purchase uranium from the Belgian Congo; Sengier told him he need look no further than Staten Island. He took out of his desk drawer a pad of lined, yellow paper, and wrote out there and then an agreement assigning the uranium to the United States Government, the financial terms to be agreed at some future time.

Later, when the US Government wanted more uranium from the Belgian Congo, talks were held in London between Sengier, the Belgian Government in Exile, and British and American representatives (30 per cent of the stock in Union Minière was British-owned). As a result, the Shinkalobwe mine, which had been flooded, was reopened, and mining was started again, and 3,000 tons of uranium ore was shipped to America from West African ports.

Sengier had played a statesman's role, and he was rewarded in appropriate coinage. He was awarded the US Legion of Merit, and knighted by the British Government. He was in Washington on V-J Day, and General Leslie Groves, the director of the Manhattan Project, took him to the White House, and introduced him to President Truman as a man without whom the atomic bomb could not have been built.

There was also uranium in the Colorado Plateau, in those sandstone conglomerates that had accumulated as the seas ebbed away 400 million years ago. Radium mining having become unprofitable there, the carnotite was being mined for vanadium by the US Vanadium Corporation, an affiliate of Union Carbide. The residue from vanadium mining contained uranium. The Army and Union Carbide jointly built a refinery and refined uranium from the tailings, though Union Carbide was not told what the uranium was wanted for.

The uranium in the Canadian North was to prove somewhat more controversial. Gilbert Labine's Eldorado Company closed down the Great Bear Lake mine during the war because it could not get the manpower or equipment to maintain it, and the mine was flooded. The US Office of Research and Development told LaBine that it wanted uranium from his company's mine. It also asked the help of the Canadian Government. C.D. Howe, the Minister of Munitions and Supply, offered to get Eldorado priority in obtaining equipment if it would reopen the mine, and mining was resumed there in August 1942. However, more uranium was acquired by combing the residue of ore that had been discarded at the Port Hope refinery after the radium was extracted. At General Groves' direction, the US Army concluded a contract with Eldorado for the purchase of 700 tons of refined uranium oxide.

However, others also were interested in Canada's uranium. The British atomic bomb project was not yet fully merged with the American one, and when the projects were joined, the British were not happy about the degree of control that they exercised over it. They wanted some Canadian uranium of their own, and they also wanted some to go to the Montreal branch of the project, which was working on the design of a nuclear reactor and the handling of radioactive materials. Some officials expressed concern that what amounted to Canada's total probable production for the next two or three years had already been sold to America.

This dissatisfaction was aired in Ottawa, and someone told C.D. Howe that Winston Churchill felt that he was 'selling the British Empire down the river.' As it happened, Howe was sensitive to the charge that he was insufficiently patriotic, since he was not born a Canadian but an American, and had moved to Canada as a young man. He was angry and upset, and talked about uranium to Chalmers Mackenzie of the Canadian National Research Council. Mackenzie went to see General Groves in New York, and told him that Howe was under fire for allowing all Canada's uranium to go to America. Groves told him that the American programme was tight and could not spare any uranium. Mackenzie said, in what must have sounded like a velvet-gloved threat, that the Canadian Government could easily overrule Eldorado's contract with the US Army on the grounds of wartime necessity, though of course it would not dream of doing so. The meeting ended with an agreement to explore jointly the possibility of stepping up the output of the Great Bear Lake mine. In fact, the Canadian Government soon decided that such a uniquely strategic resource as uranium could not be left in wartime in the hands of a private company, and nationalized Eldorado. The Government-owned

Eldorado Mining Company is still a major producer of uranium today.

In the event, two-thirds of the uranium used for the wartime atom bombs came from the Congo. The rest came almost equally from the Great Bear Lake and the Colorado Plateau vanadium mines.

In August 1943, General Groves commissioned an evaluation of the world's uranium resources. The conclusion was that there was enough uranium available for the Manhattan Project, but after this, there would hardly be any left in North America, and the only source of supply would be the Belgian Congo. This evaluation was presented to President Roosevelt. The illusion that uranium was a scarce resource was to persist for some time, and was to be an important determinant in nuclear power policy.

* * *

It is difficult for generations raised in a world of nuclear arsenals and the ever-present possibility of military annihilation to appreciate the impact that the atomic bombs dropped on Hiroshima and Nagasaki made upon people's thinking. The world had just been through a long war that seemed as terrible as a war could be in which hundreds of thousands of bombs had been rained down upon cities. Now, suddenly, *one single bomb* could destroy half a city. Another war, which seemed almost unthinkable after five years of murderous world conflict, now seemed to be impossible. The vision of a future war, insofar as it could be conceived, was something like that of a chess game in which every piece could take all the other pieces at once and kick over the board as well; it was a game that could not be played. And yet there was the awful fear that it might be. The bright dawn of the postwar world had an ominous dark cloud in it, and it was mushroom shaped.

A bestselling book with a widely acceptable message was called *One World Or None*. Another lengthy and much-quoted essay was called *Modern Man Is Obsolete*. This reflected the depth of the widespread feeling that everything was changed with the appearance of the atomic bomb, that international relations, war and peace, nations and sovereignty, no longer had the same meaning that they had throughout history.

This feeling was shared by some of those in power. The British

Prime Minister, Clement Attlee, was by nature a pragmatic parliamentarian rather than a prophet of the Apocalypse, but he was moved to write to President Truman suggesting that they consult with Russia on outlawing war for all time. 'The time is short,' he wrote. 'I believe that only a bold step can save civilization. If Mankind continues to make atomic bombs without changing the political relationship between states, sooner or later these bombs will be used for mutual annihilation.'

President Truman outlined the problem created by the invention of the atomic bomb in a speech in October 1945, two months after the war's end, and it is hard to fault it today. He made several basic points about atomic energy. The first was that no nation can keep or morally defend a monopoly on the peaceful benefits of atomic energy. The third was that there is no defence against atomic weapons, and the fourth was this: 'All the initial processes in the production of fissionable materials, and certain subsequent processes, are identical, whether their intended use or purpose is peaceful or military.' This was a thought that was to echo down through the years and echoes still.

For many people, the idea that the nation-state was no longer sovereign was implicit in the creation of the United Nations. The organization was seen widely as a world government in embryo. To the moral will to create a world law that lay behind many people's hopes for the United Nations, the existence of the atomic bomb seemed to add the practical necessity for it. When the UN General Assembly held its first meeting, in January 1946 in London, resolution number one called for the setting up of an Atomic Energy Commission. The resolution looked ahead to international control. It instructed the Commission to 'make specific proposals for extending between all nations the exchange of basic scientific information for peaceful ends; for control of atomic energy to the extent necessary to secure its use only for peaceful purposes; and . . . for effective safeguards by way of inspection and other means to protect compliant states against the hazards of violations and evasions.'

Any proposals on atomic power that counted for anything were going to come from the only country that actually had atomic power. Just before the UN General Assembly met, a committee was set up by President Truman to draft a plan for the international control of atomic energy, headed by Dean Acheson, then Under-

Secretary of State, and David Lilienthal, and including a number of eminent scientists who had worked on the bomb. Neither Acheson nor Lilienthal were unworldly men. Lilienthal was spurred on in public life by the idealism characteristic of the New Dealers, but he had much experience of practical politics as the head of the Tennessee Valley Authority. Acheson, who was to be one of the creators of the Western alliance, was a very hard-headed statesman, who often expressed urbane contempt for expressions of lofty principles. Yet the proposal they drew up for the control of atomic energy was more sweeping in its nature and more generous in spirit than any that has been put forward since, offering to surrender national sovereignty in vital areas to a new international body. That these two men could have put forward such a plan shows how drastic a change in the world situation was perceived. It addressed itself to the heart of the problem, so that subsequent plans to bottle the genie of atomic power can be judged by how much of the ground of this plan they covered.

The Acheson-Lilienthal Plan, as this came to be known, proposed the creation of an International Atomic Development Agency, to which all raw materials and atomic facilities in the world would be entrusted. This agency would have the right to license and control all activities in any country, and ensure that they were for peaceful purposes only. It would have the duty of 'fostering the peaceful uses of atomic energy.' Once the régime was set up, then all atomic weapons would be destroyed, and the manufacture of atomic bombs would stop. This would be a true supra-national régime, because it could impose sanctions for the violation of its rules, and these would not be subject to a veto, like measures of the UN Security Council.

This American plan would remove from American control the laboratories and plants where the atomic bomb was developed and built at great expense. It would also bring under the Authority all the uranium deposits in the world.

The plan was presented to the UN Atomic Energy Commission by Bernard Baruch, with a speech in which he pulled out all the oratorical stops. 'We are here to make a choice,' he said, 'between the quick and the dead. Behind the black portent of a new atomic age lies a hope which, seized upon with faith, can work out our salvation. If we have failed, then we have doomed every man to be the slave of fear.'

There was one difference of detail between the plan that Acheson and Lilienthal presented to the State Department and Baruch's presentation at the United Nations, and this was an issue that was to dog control of atomic energy for years to come. This concerned so-called 'denatured' material, which meant somehow contaminating fissile material so that it could not be used as an explosive.

The State Department report of the plan in March said: 'Uranium and plutonium . . . may be denatured. Efforts to make atomic bombs with denatured fissionable material have failed in the United States. . . . If only denatured material were issued, control would be feasible.'

This came as a surprise to a number of scientists, and set off discussion. By the time Baruch presented the plan to the UN Commission, he found it necessary to issue a cautionary note: 'Denatured materials, whose use we know also requires suitable safeguards, should be furnished for such purposes by the Authority under lease or other arrangements. Denaturing seems to have been over-estimated by the public as a safety measure.'

The Acheson-Lilienthal Plan was welcomed widely as a revolutionary response to a revolutionary situation. There were some misgivings, it is true. The British Government, already aggrieved because it was cut off abruptly from the post-war benefits of its wartime partnership with the United States in atomic energy, had plans to push ahead with both civil and military nuclear energy, and it was worried that this might hobble them. Four members of the UN Atomic Energy Commission, rich in primary resources, forced a slight modification of the plan, so that uranium would come under international control only after it was mined. Some members of Congress were very unhappy at the thought of turning over America's nuclear installations to a supra-national authority ('globaloney' was the term used to deride such internationalism).

The Russians rejected the plan. Their Marxist view of a world divided into inevitably hostile camps did not admit the idea of a common issue that could override the interests of both. They put forward some practical objections: they would have to surrender their uranium deposits and nuclear laboratories while America still retained its atomic bombs, which were to be destroyed at a later, unspecified date. They were also worried about the normal pro-American majority in the United Nations, and about placing what was potentially an important new source of energy under its control. They put forward an alternative plan that would require the United States to destroy all its atomic weapons before the supra-national body was set up. This was clearly unacceptable to the Western powers, and the Acheson-Lilienthal Plan was rejected.

The nearest thing to the projected Authority that has been created is the International Atomic Energy Agency. This has some of the powers of inspection that were envisaged for the Authority, though not of control; it is a more limited attempt to deal with the same problems of the overlapping of nuclear power with nuclear weapons, and of the demands of national sovereignty and global security.

Chapter 2

FINDING URANIUM AND USING IT

A uranium fission chain reaction can be made to produce power explosively, in a bomb, or else slowly, in a nuclear reactor. Both were done during the building of the first atomic bombs. In fact, everything important that has been done with uranium was done during those years. The thermonuclear or hydrogen bomb, which uses the fission bomb as a trigger, was still to come, but that is another story.

The first uranium chain reaction in 1942 was made with the same materials that many nuclear reactors use today: uranium, in blocks, and graphite, interspersed in a lattice as a moderator, to slow down the neutrons. It was an experimental reactor, built to see whether a chain reaction could be achieved; it generated 200 watts, enough to light two household light bulbs, and it ran for three months. The next uranium pile, as these are called, built at Oak Ridge, Tennessee, was much like one of today's reactors laid on its side; a cube of graphite containing horizontal channels, filled with uranium metal rods encased in aluminium. It generated 800 kilowatts (800,000 watts).

Only two substances have been turned into a fission explosive, so far as we know: highly enriched uranium, which is mostly u-235, and plutonium, which is a product of uranium fission. Uranium 233, which is even more rare than u-235 though it is also produced by certain fission reactions, can be made into an explosive too, but no one seems to have done it. Enriched uranium, not so highly enriched as that used in bombs, is the fissile material for most reactors today; however, some use natural uranium, that is not enriched.

Enriching uranium means increasing the percentage of u-235 by a winnowing process that removes a lot of the u-238, which makes up more than 99 per cent of it. This proved to be by far and away the most difficult single task of the wartime atomic bomb project. Another term for it is isotopic separation. Ways have been devised to separate isotopes by using the mass of the atoms, a delicate process since the mass is so slight, and the difference made by the addition or subtraction of one or more neutrons slighter still. The heavier the atom, the less the difference in mass of an isotope and hence the more difficult it is to separate it. In the case of uranium, the difference is very slight indeed, since an atom of ordinary uranium has 146 neutrons and 92 protons in its nucleus, and one of u-235 has 143 neutrons and 92 protons. (The number of protons in the nucleus determines the atomic number that is assigned to an element, the number of protons and neutrons its atomic weight. Thus, uranium is element number 92, the highest in the table of natural elements, and its atomic weight is 233, 235 or 238.)

The atomic reactors built during wartime used natural uranium, but the most widely used reactors since then, the pressurized water reactors built and sold by the United States, use enriched uranium as fuel. Since only the u-235 atoms undergo fission, fuel use is more efficient if the uranium is enriched. For fuel, the uranium is enriched to between 3 and 3½ per cent. For an atomic bomb, the uranium should be enriched to about 95 per cent and the minimum that will work must be somewhere above 50 per cent. The degree of enrichment is crucial. The difference between 3½ per cent enriched uranium and 90 per cent enriched uranium is the difference between something that can provide heat and light to a city, and something that can blow it off the face of the Earth.

The word 'enrich' was applied to this process during the Manhattan Project days, and it has stuck, but it is a curious word to use for it. It is erroneous. It is also significantly and even dangerously misleading. It gives a false impression of the vital difference between uranium enriched for power and uranium enriched for explosives.

In the common dictionary definition, and in everyday use, enriching a substance means adding something to it to give it a higher quality. But nothing is added to the uranium. It does not receive an injection of u-235. The proportion of u-235 is increased, not by addition, but by subtraction, by removing some of the u-238.

To see the importance of this distinction, try to answer this question quickly: if someone has an enrichment plant, and has enriched uranium from its normal 0.7 per cent u-235 to 3½ per cent in order to use it as reactor fuel, how much further does he have to go to enrich it to 90 per cent, so that he can make atomic bombs with it? The snap answer might be that he has taken only the first small step, and has a long way still to go. But this is not so.

It would be true if he were really enriching it. Say you start with 100 lbs of uranium. Seven tenths of one pound of it is U-235. If you were adding more U-235, then to get it up to $3\frac{1}{2}$ per cent, you would add slightly more than 2.8 lbs. Then you would have 102.8 lbs of uranium that is about $3\frac{1}{2}$ per cent U-235. If, then, you wanted to add enough to make it 90 per cent U-235, you would have to add just over 800 lbs of U-235. Then you would have 1,102.8 lbs of uranium, of which 902.8 lbs would be U-235, or fractionally over 90 per cent. In the first stage, enriching uranium for fuel, you would have gone about one-three hundredth of the way to enriching it for a bomb.

But you are taking matter away, not adding it. To reduce that 100 lbs of uranium so that it is $3\frac{1}{2}$ per cent U-235 instead of 0.7 per cent, you will separate out and remove 80 lbs of the U-238. To get to 90 per cent U-235, you just take away 19 lbs more. You will have done most of the work already.

One might think that in practice, the law of diminishing returns would operate heavily against you by that time, but it turns out that it does not. A scientist who has been working on uranium enrichment for thirty years, Dr Karl Cohen of Exxon Nuclear, told the US Nuclear Regulatory Commission in testimony that it takes one-quarter as much enrichment work to produce bomb-grade uranium from $3\frac{1}{2}$ per cent enriched fuel as it does to produce it starting with natural uranium. So a person who has enriched uranium for fuel has gone three-quarters of the way towards making the material for an atomic bomb.

This is why an enrichment plant is such a sensitive facility. This is why a country's decision to build a uranium enrichment plant so that it can make its own reactor fuel instead of having to import it is a strategic decision as much as an economic one, or at least, is likely to be seen that way by that country's neighbours.

The most common method of enriching uranium is by gaseous diffusion. The uranium is turned into a gas, uranium hexafluoride, and pumped through a porous membrane. More of the lighter gas than the heavier gas goes through, so that the gas on the other side of the membrane contains a slightly larger proportion of U-235 atoms. The gas is passed through membranes again and again, leaving some u-238 behind each time. The principle is easy to understand, but the practice is very difficult. Pumping the gas around requires so much power, the membranes have to be so fine, and the materials kept at such a degree of purity, that a gaseous diffusion plant approaches a moon rocket in its engineering requirements and even its cost. For a while, the uranium enrichment plant at Oak Ridge, Tennessee used more than 10 per cent of the electricity supply of the entire United States, and even in 1963, enrichment plants were using 7 per cent of the country's electric power output. Only a technically advanced and rich country, or a spendthrift one, can enrich uranium by gaseous diffusion. However, a more efficient way of enriching uranium has been devised, first of all in Europe, the centrifuge method, though this only cuts the cost drastically if it is operated on a very large scale.

So far, there are not many enrichment plants in existence. The five nuclear weapons powers have them. There are four others in Western Europe, all under multi-national ownership, and a fifth under construction. South Africa and Japan have small ones, and Brazil is planning one.

* * *

There is another substance besides u-235 that can be used to make an atomic bomb. Back in 1940, a team of scientists at the University of California at Berkeley under Glenn Seaborg created a new element, a 93rd, by bombarding uranium with neutrons. Some of the neutrons were captured by uranium atoms. What happened then involves a change, not only in the atom, but in the neutron also, for the neutron, unlike other atomic particles, is divisible. The captured neutron lodges in the nucleus, but it emits a small bit of itself in the form of an electron. It then has a positive charge, since it was electrically neutral before and emitted a negatively charged particle. With a positive charge, it is no longer a neutron but a proton. The element now has 93 protons instead of 92, so it is no longer uranium but a new, 93rd element. Seaborg called this neptunium.

Neptunium does not last long. It is radioactive, and has a half-life of fifty-three minutes. However, the nuclei of some neptunium atoms capture another neutron, and the same thing happens: the neutron emits an electron and turns into a proton. So the atom now has 94 protons. Seaborg had created another new element, not found in nature, a 94th. He called it plutonium. This also is radioactive, but it has a half-life of 24,400 years. It can fission and produce a chain reaction, just like uranium 235.

Ernest Lawrence, who had earned a Nobel prize in physics for his invention of new machines to investigate atomic structure, was also at Berkeley, and he was associated with the Manhattan Project. A chronically enthusiastic giant of a man, he saw new possibilities in the fact that plutonium, a fissile material, could be manufactured from uranium. In July 1941, he wrote in a memorandum:

> If a chain reaction with unseparated isotopes is achieved, it may be allowed to proceed violently for a period of time for the express purpose of manufacturing element 94 in substantial amounts. This material could be extracted by ordinary chemistry. . . . If large amounts of element 94 were available, it is likely that a chain reaction with fast neutrons would be produced. In such a reaction, the energy would be released at an explosive rate which might be described as a 'super-bomb'.

His advice was followed to the letter. A reactor was built at Oak Ridge, Tennessee, and then two others at Hanford, Washington, in which the chain reaction would produce plutonium. Then the plutonium was separated chemically from the rest of the metal, just as Lawrence had proposed. Because it was wartime and there was an urgency about building the bomb, the u-235 programme and the plutonium programme were carried on in parallel, instead of one being tried to see whether it worked before the other was tried. By the summer of 1945, enough plutonium was created to produce two nuclear bombs, and enough uranium 235 to make one. The first atomic bomb, tested at Alamagordo, New Mexico, in June was a plutonium bomb. The bomb dropped on Hiroshima was a u-235 bomb. The second plutonium bomb was dropped on Nagasaki. In principle, the mechanism for producing an explosion with each material is the same: a quantity of the material sufficient to make up a critical mass is brought together very suddenly, so that an explosive chain reaction takes place.

The Acheson-Lilienthal Plan spoke about 'denatured' fissile material, that cannot be fashioned into nuclear bombs. In the case of uranium, they obviously meant low-enriched uranium. In the case of plutonium, the situation is a little more complicated. We have seen how, in the fission process, some u-238 turns into plutonium. This is plutonium 239. But the change does not stop there. As the fission process continues, some atoms of plutonium 239 capture one or two more neutrons to become plutonium 240, 241 or 242. Plutonium 240 and 242 are not fissile. In making the plutonium for the first atomic bomb, the uranium fuel rods were taken out of the reactor as soon as some of the plutonium 239 was formed in it, and before much of it could be turned into the higher isotopes. Evidently, at the time that the Acheson-Lilienthal plan was drawn up, scientists believed that if the fuel rods were left in the reactor longer, so that most of the plutonium is 240 or 242 rather than 239, then this could not be made into an atomic bomb, and so would be denatured. The situation is complicated, but suffice it to say for the moment that they turned out to be wrong.

Plutonium is important because it not only *can* be produced by the nuclear fission of uranium, but it is *always* produced by it. It has become central to the relationship between nuclear power and nuclear weapons, inevitably; it is an output of the first and an input of the second.

It has acquired a fearful image. Some people who should know better have said that plutonium is the most toxic substance known to Man. This is the kind of statement that is high on drama but low on precision. You can hold a piece of plutonium in your hand without any ill-effects, you can put a match to it and it won't explode, and you can even sprinkle some on a sandwich and eat it and probably get away without any ill-effects. This makes it in different ways less harmful than radium, dynamite or strychnine.

Plutonium is radioactive, and its radioactivity is in the form of alpha radiation. This consists of a package of two protons and two neutrons, and these are by far the heaviest kind of radioactive emission. (Beta radiation consists of single electrons, and gamma radiation is like x-rays only with a shorter wavelength and more powerful.) When any of these forms of radiation encounter matter, they knock away one or more electrons from the atoms they hit, turning these atoms into positively-charged ions, and so the radiation is called ionizing radiation. The heavier the radioactive emission, the slower it travels, and the easier it is to stop. Alpha emissions travel relatively slowly, have a very short range and very little penetrating power. They can be stopped by a thin sheet of paper, the outer layer of the skin, or even the lining of the stomach. This is why you can hold the plutonium in your hand without its radiation penetrating your skin, and why if you ate it, the radiation would not penetrate the lining of your stomach as it passed through your digestive system. It is dangerous to human beings in only one respect: if some particles of it are breathed in, even a minute quantity, then their radiation will attack the walls of the lungs with a high probability of causing cancer eventually. Even a few particles of plutonium dust in the air are very dangerous.

When the plutonium is produced in a reactor it is mixed with other products of fission, plus the uranium 238, in a metal fuel rod that is very hot and radioactive. The plutonium must be extracted from this. As Lawrence observed in his memorandum to the heads of the atomic bomb project, this can be done by 'ordinary chemistry', but nonetheless, it is delicate work, and dangerous also. Rigorous safety precautions must be taken, so that no plutonium dust gets into the air. But the task of plutonium extraction is well within the capabilities of any country with a fairly advanced technology.

One last word about atoms before we leave their world for a while. The familiar model of the atom as a collection of particles orbitting around other particles is a useful one; physicists employ it all the time, and often think of atoms in this way. But this is not a representational model, in the way that a model airplane may be a scale model of the full-sized aircraft flying in the sky. It is more like the model that an economist concocts of a free market in equilibrium, or a psychoanalyst of the layers of super-ego, ego and id: a diagramatic illustration to represent something that is really abstract, a metaphor rather than a picture. The atom, if it could be magnified a few billion times so that it would be visible, would not *look* like this. For one thing, it can be demonstrated that, in certain circumstances, an atom behaves in a way that is incompatible with this model, as an energy wave rather than a collection of particles. What, then, does an atom *really* look like? The question does not make sense in terms of modern physics. The atom is represented to the physicist by a collection of mathematical formulae.*

* Frisch wrote of Niels Bohr, who more than anyone else created our model of the atom: 'With Bohr we give up the naïve concept of reality'.

The possibility of using nuclear fission as a source of heat and light was seen from the beginning: the energy output of the reactors built during the Manhattan Project was measured in watts, just like the energy output of a power plant. It was seen that the heat could be made to produce steam and be harnessed to an electrical generator. But when this would become a practical proposition, and if it ever would, were open questions. For most people, atomic power meant the atomic bomb, and atomic fission was a source of menace rather than hope. Thus David Lilienthal, the first Chairman of the US Atomic Energy Commission, recorded in his diary on 4 August 1947, a conversation with the AEC's Director of Research:

> We considered the possibility that there wasn't a lot of uranium in the world in rich deposits, and low grades would be so expensive to handle that it wouldn't be worth the trouble. I said 'wouldn't it be wonderful if we could take what we have in the way of plutonium etc. and dump the stuff into the sea, and know that no one else would find enough ore to bother with?' We both beamed.

There was no thought that any boon to Mankind would be lost beneath the waves.

Certainly there were non-military uses for nuclear energy, particularly the production of radio isotopes for use as tracers in industry and medicine. But nuclear power was another matter. Those who offered opinions on the subject varied enormously in their judgements. An article in the *New York Times* by Lindesay Parrott in March 1946, indicated the range of forecasts: 'Some experts see atomic energy as a major source of the world's power within five to fifty years.'

A group of scientists and engineers reported to Bernard Baruch in September 1946 that a 75,000-kilowatt nuclear power plant could be built immediately at a cost of $25 million, and that it would produce electricity at a cost of 26 per cent higher than that of a coal-burning plant in the north-eastern United States. They said this assumed an operating capacity of 100 per cent and 3 per cent bank interest rate. Those were considerable assumptions. No nuclear power plant has ever reached an operating capacity of more than 85 per cent, and a 3 per cent interest rate belongs in the realms of financial daydreaming or nostalgia.

Some newspaper and magazine articles talked glowingly of the promise of nuclear power. William L. Laurence, the respected science correspondent of the *New York Times*, wrote in his book *Men and Atoms*, published in 1946, that the United States could pass on the technology of nuclear power to other countries: 'Such an atomic lend-lease could beat atomic swords into atomic plowshares, and transform the Earth into a promised land of plenty for all Mankind.'

Robert Oppenheimer, the Chairman of the AEC Advisory Committee, was more cautious. Testifying before a congressional committee, he said electricity could be produced by atomic power in five years' time, but estimated that it would be ten to twenty years before it would be competitive in industrialized countries, and thirty to forty years before there would be large-scale development. Dr Oppenheimer, so wise in many respects, under-estimated the economic and also the political appeal of 'atoms for peace'.

* * *

The post-war decade was one of a rush for uranium. When Gordon Dean retired as Chairman of the AEC, he wrote a book called *Report on the Atom.* At one point in this, he called attention to the need for uranium for the US nuclear weapons programme in a tone so urgent that it sounds almost frantic:

> The security of the free world may depend on such a simple thing as people keeping their eyes open. Every American oil man looking for 'black gold' in a foreign jungle is derelict in his duty to his country if he hasn't at least mastered the basic information on the geology of uranium. And the same applies to every mountain climber, every big game hunter, and, for that matter, every butterfly catcher.

In fact, many mountain climbers and butterfly catchers did master the basic geology of uranium, and carried pictures of telltale streaks in their minds and geiger counters in their pockets as they climbed cliff faces or pursued lepidoptera.

Governments set off the uranium hunt. The Combined Policy Committee, in which Britain, the United States and Canada were supposed to decide policy on atomic energy, was dissolved after the war, but the organization it had established to find and acquire raw materials, the Combined Development Trust, was kept in being. This was partly to ensure that the allies did not compete against each other for materials, and also partly because much of the globe was a part of the British Empire still, and the United States thought Britain could be useful in some areas. The Trust changed its name to the Combined Development Agency in 1948 on the advice of Edgar Sengier, whose voice was still heeded, and who suggested sagely that the word 'trust' might convey abroad a sinister suggestion of a commercial monopoly.

In 1946, the three governments agreed, after some wrangling, to divide up the Congo uranium among them according to a complicated formula, with the most going to America on the ground that America had the most pressing need, for its weapons programme. The Belgian Government agreed to this arrangement, and did not set a high price. It was felt in Washington and London that this was partly an expression of Belgian gratitude to the countries that had liberated her from German occupation, and later, that it was a part of Belgium's contribution to the defence of the West.

The Combined Development Agency reached an agreement with the Australian Government to buy all the uranium found in Australia and to co-operate in exploration. It had been assumed that Britain would conduct all negotiations on behalf of the CDA with its Empire partners, but the Australians preferred to negotiate with the United States, a reminder of the changed power relationship in the Pacific.

Uranium was found in South Africa. An American geologist, George Bain, became a consultant on the Manhattan Project and, fired by this new interest, he tested every rock specimen he could find with a geiger counter. He found traces of uranium in one that he had brought back from a field trip to the Wittwatersrand gold fields. It turned out that the waters that, two billion years ago, carried specks of gold along with particles of quartz and other minerals to form conglomerates had carried specks of uranium also. The CDA approached the South African Government with a view to exploiting this find. Agreement was near when the Smuts Government was voted out of office, beginning an apparently unending period of rule by the National Party, which was less friendly to Britain. An agreement to supply uranium was reached finally in 1950: South Africa was to receive technical assistance in the exploitation of atomic energy as part of the *quid pro quo*. Mills were set up, and the ore from which gold had already been extracted was combed for uranium. Later, it was found to be easier to extract the uranium first.

Portugal was seen as an important source of uranium; there were radium mines there, and other deposits seemed likely to be found. Since the mines were partly British-owned, the British ambassador in Lisbon, Sir Nigel Ronald, was given the mission of asking the Portuguese Government to give the CDA a monopoly of its uranium. Sir Nigel saw Dr Salazar, the ascetic, right-wing academic who had been Portugal's dictator for twenty years, and explained the workings of the CDA. Salazar remarked wryly that the decision by Britain and the United States to divide up the world's uranium between them reminded him of Pope Alexander Borgia dividing the whole of the New World between Spain and Portugal. Sir Nigel said that by allotting the uranium to the CDA, Portugal would be contributing to world order, no doubt sensing that 'order' was a good word to use to Salazar. Salazar said he wanted more tangible benefits for his country. Eventually, terms were settled, but the uranium deposits proved smaller than was hoped, and Portugal as a source of uranium less important.

In 1948, uranium in useful quantities was discovered in pitchblende in France, near Limoges, and soon afterwards in two of France's African colonies, Niger and Gabon. France now had enough uranium for her immediate needs, and she continued to exercise control over the mines in Niger and Gabon after those countries were given independence.

It turned out that there were far more uranium deposits in the world than had been supposed. The variety of uranium occurrences continued to complicate exploration and to make forecasting very uncertain. There have been big new discoveries in America and Canada, and Australia also. A big deposit was found in South-West Africa, now Namibia, in 1966, when Peter Louw, a lifelong mineral prospector, persuaded a mining company to follow up the discovery of radium-bearing pitchblende that he had made in 1928, in a stretch of grey, rocky desert.

The early 1950s were the heroic days of uranium exploration, the days of the lone prospector dreaming of a pot of uranium at the end of the rainbow that would make him a millionaire, and occasionally finding it.

In the United States, the Atomic Energy Commission set out to stimulate exploration. Under the 1946 Atomic Energy Act, only the AEC could possess fissile material once it had been mined. The AEC offered to buy uranium ore that was better than 0.1 per cent grade, and to pay $13 a pound for the uranium. Its Division of Raw Materials opened offices in Colorado, Utah and New Mexico, advised prospectors, and put out a booklet called *Prospecting for Uranium*. It also opened mills to extract the uranium from the ore.

Uranium fever swept the Rocky Mountain states. *Prospecting for Uranium* sold one hundred thousand copies. The geiger counter, which detects radioactivity, had until then been a specialized instrument, but now, in the West, it was sold at hardware stores, sporting goods stores and even drug stores, and in 1952 it was listed in the Sears Roebuck mail order catalogue. The bestselling geiger counter was the Super Sniffer, produced by the Nuclear Chemical and Instrument Corporation in Chicago, which retailed at $49.50. People going into the countryside for a picnic would take a geiger counter along and pass it over the ground just in case. Some people became weekend prospectors; others tried it as a full-time occupation, sometimes the same people who prospected for other metals.

Many people found uranium. Most of the deposits were small, and on the surface, typically in soft sandstone rock. One such would be mined by half a dozen men, who would break up the ore with a jack hammer, load it into a truck and drive it to the nearest AEC mill. In 1954, there were five hundred working uranium mines. Shares in uranium mines were sold in bars, barber shops and ladies' bridge clubs.

Some people struck it lucky. One of the first was Paddy Martinez, an illiterate Indian sheep-herder who lived near Grants, New Mexico. In 1950, he was riding to a trading post to buy some cigarettes when he saw a yellow spot under a rock. He dug it out with a stick because it reminded him of a piece of uranium ore he had seen once in Grants, in the hands of someone who was saying that this kind of thing was worth money. He got a local farmer friend to fill in the claim form, then sold the claim to a mining company in exchange for an income. (The AEC found that Navajo Indians were excellent uranium hunters because of their memories of the landscape. A Navajo would sometimes be shown a picture of a uranium seam in a rock, and remember where he had seen rocks with this colouring ten years earlier.)

Charles Steem was the first uranium millionaire. He was a Texan who had worked as an oil geologist, and had a theory about how uranium could be found. He spent all his money renting a drill rig and went prospecting in Utah. He and his family were at starvation level when he finally found uranium. He formed the Utex Corporation to manage the mine, and made $2\frac{1}{2}$ million in its first year of operation.

He became a legend throughout the West. Many a uranium hunter struggling against poverty and discouragement would cheer himself up by recalling how a small-town bank in Colorado had refused to lend Steem $250 to buy drilling equipment, and how Steem got his own back later by buying the bank. Steem launched himself on a millionaire life style; he bought a ship from the British Navy as a yacht, built himself a palace in Las Vegas, and flew friends in from all over the United States for parties several times a week. He branched out into other business ventures disastrously, and died owing money to the Internal Revenue.

Vernon Pick's find was also dramatic, but his story has a happier ending. After his machine shop in Royalton, Minnesota, burned down, he and his wife bought a panel truck and trailer and went on a long vacation in the south-west. In Colorado Springs, Pick met people hunting for uranium and was bitten by the bug. He was forty-seven. Nine months later, he was broke and weak from lack of proper food. He rested by a stream while hiking down a hillside after a fruitless search, when his geiger counter started clattering. He staked a claim, and borrowed the money on his truck and trailer to rent a bulldozer and a jack hammer. He made $400,000 from the mine in two years, and then sold it to the Atlas Mining Corporation for nine million dollars. He moved to California and started an aerial exploration company, which he ran successfully.

A few others struck it rich. Bob Schwartzwalder was the janitor of a school in Golden, Colorado, and a long-time 'rock hound' on the weekends, when he would go out looking for interesting rock specimens. When he took a geiger counter along, he found uranium, formed the Schwarzwalder Mining Company, and sold it to Consolidated Edison. Robert Adams owned a small restaurant in

Rawlis, Wyoming; he hired a young pilot to fly his Cessna low over the countryside with a scintillometer, a more sensitive detector of radiation than a geiger counter, and throw out a sack of flour whenever it registered. He found a big deposit at one white smudge, formed the Energy Fuels Corporation in Denver, sold it for more than a million dollars, and is still in business in Denver now, in uranium and coal.

The big mining companies were not interested in uranium at first because early experience indicated that it was to be found only in small deposits, which would not warrant large-scale investment. However, Kerr-McGee found what turned out to be the largest uranium deposit in America, at Ambrosia Lake, New Mexico, when it was looking for oil.

By 1958, the AEC had all the uranium it needed for weapons and research. It ended the great uranium hunt by announcing that it would buy no more uranium from new deposits.

In Canada, the search for uranium took a different turn. There were fewer discoveries, but bigger ones. Eldorado continued to mine uranium at the Great Bear Lake. Then it sent exploration teams to Lake Abathasca, in northern Saskatchewan, where pitchblende deposits were known to exist, and found useful quantities of uranium there. The Government opened up the area to claims, and dozens were staked by individual prospectors, many of them profitably. The Federal Government laid out a townsite and named it Uranium City, and it is a flourishing community still.

The biggest uranium strike was in the forests of northern Ontario, and this was the work of individual prospectors. Karl Gunterman was a German immigrant in Sault Sainte Marie, and he tried his luck at uranium prospecting in partnership with Aime Breton, who owned a small hotel in the town. In 1949, they found a piece of rock in a mining office that made a geiger counter click, established that it came from Long Township, nearby in the maple and fir woods of northern Ontario, and staked a claim on the site. Then the pair met Franc Joubin, a geologist who had played a part in the Lake Abathasca find, and feeling the need of some expertise, they invited him to buy a share in their claim. He went to see it, found quartz pebble conglomerates there, and got strong responses from the geiger counter. He collected some samples of the rock and sent them to the Bureau of Mines office for an assay. The result, to his surprise, was negative. He decided that the radioactivity must be coming from thorium, and told Gunterman and Breton that their claim seemed to be worthless. Under Canadian law, a claim must be worked for forty days a year if it is to be retained. Gunterman and Breton let theirs lapse.

Later, on a trip to London, Joubin talked to C.F. Davidson, a British geologist who had been involved in the CDA's search for uranium in South Africa, and learned from him that where uranium is associated with pyrite and quartz, it may be leached away from the surface by water. He remembered Long Township, and decided that there might indeed be uranium there, but deep down, the uranium on the surface having been leached away, which would explain why the assay was negative. This could only be established by drilling. In 1953, he staked his own claim to the area, and got financial backing for a drilling operation from the American financier Samuel Hirshhorn, for whom he had worked on another mining project. They drilled deep down and found a rich uranium deposit that extended along the shores of Elliot Lake. They formed a company to exploit this find, and, using dozens of Hirshhorn's employees as agents, they mounted a secret operation and staked claims on the same day to sites covering more than 20 square miles. The shares in their company shot up from one dollar to $125 each. Joubin made more than a million dollars from the find, and Hirshhorn made much more. Joubin stayed in prospecting and mining, worked in several countries, and at this writing, a sprightly, silver-haired man in his seventies, he is a consultant to two UN agencies. Hirshhorn later sold his share in the mine to Rio Algom for $81 million and created the Hirshhorn Museum in Washington DC, with its famous collection of modern painting and sculpture. Aime Breton died a bitter man, telling friends he had been cheated out of a fortune.

These finds were followed by others around Quirke Lake, seven miles away. The quartz conglomerate, it turned out, was in a long vein underground, extending many miles. The principal prospector at Quirke Lake was one Carl Mattaini, then the head of a small airline called Lau-Goma Airways, but he later brought in the huge Consolidated Denison Mining Company.

Long shafts were dug underground at Elliot Lake and Quirke Lake before the infrastructure of community grew up above ground. The miners lived in tents at first, had their food flown in and cooked at field kitchens, and drove along a dirt road to the little town of Blind River on Saturday night to drink. Men came to Elliot Lake from all over the world for the high wages, and the opportunities also. For a while it was a rough, brawling frontier town, with a lot of hard drinking, fights on Saturday nights, and a few shootings. Nonetheless, by 1959 three-quarters of the uranium mined in Canada was coming from the Elliot Lake area.

When uranium was found in those days, the mining men went in before the scientists. The scientists could have told them something about uranium relevant to mining it, and some scientists did tell them, but to no effect. After uranium decays by radioactive emission into radium, the radium decays into radon, a radioactive gas. So where there is a deposit of uranium, there is also radon gas. These days, miners are limited strictly to the amount of time that they can spend in the presence of radon gas, and radon levels in mines are monitored constantly. In underground mines, huge pumps keep air blowing through the tunnels continuously to disperse the radon, so that there is a constant breeze. In some mines in south-west America the sandstone is porous and retains a lot of radon gas, so when it is blasted with explosives prior to sifting for uranium, the mine is evacuated for a few days afterwards to allow the concentration of radon to reduce.

But in the early days of uranium mining, side effects were overlooked, and no such precautions were taken for miners. Some of the consequences of this oversight were aired before a Senate public health subcomittee chaired by Senator Edward Kennedy in the summer of 1979. Widows of miners who had died of lung cancer testified before the committee that in the 1950s, after their husbands came home from work, geiger counters they kept at home showed that their bodies were radioactive.

Others told of a number of cases of lung cancer among Navajo Indians who had been uranium miners in New Mexico. Dr Vali Spinoza, formerly of the Indian Health Service, told the committee: 'Exposure was so intense in those early mines . . . it was like lying under an x-ray machine day and night for six months.' And Duncan Holaday, formerly of the Public Health Service, said: 'We in the PHS did medical exams, put out brochures; we had meetings with mine operators. They didn't believe one doggoned word we said.'

The actual consequences of ignoring radon gas were less dramatic than this testimony would imply, but they are real and measurable. A recent study by a National Academy of Sciences committee found a greater incidence of lung cancer among men who worked in uranium mines before controls on exposure were instituted. It calculated that each year that a man worked in the mines then increased his chance of contracting the disease by .003 per cent. A Canadian study produced more significant findings still. Among eight thousand men who worked in uranium mines in Ontario in the 1950s and '60s, forty-one died of lung cancer, whereas the average among a group of eight thousand men is thirteen.

One reason that the issue was not discussed more publicly at the time was that the AEC was still engaged mostly on weapons work, and had acquired habits of secrecy. Another subject aired at the 1979 Senate subcommittee hearings was the spread of radiation in nuclear weapons tests in Nevada in the 1950s. Much of this became public for the first time.

So little consciousness was there of the significance of radon gas when uranium mining began, that in Grand Junction, Colorado, tailings from a nearby uranium mine were mixed with cement and used to build houses. Twenty years later, these were found to be slightly radioactive, with radon gas retained in the basement, and the houses were pulled down and the owners compensated by the Federal Government. In Port Hope, Ontario, they had built a schoolhouse with tailings from the uranium refinery. Residents were horrified when slight radioactivity was detected in the classrooms, and the building was demolished.

The Australian Government, like the American, set out to stimulate uranium exploration after 1945. The Australian Atomic Energy Commission and the Bureau of Resources offered rewards for finds, distributed pamphlets on how to look for uranium, and gave out a stream of information on results of serial surveys. Australia also went through a period of uranium fever in the 1950s, with prospectors roaming the countryside and a boom in geiger counter sales. Mineral mining has always been a big industry in Australia, and as in the Rocky Mountain states in America, there is a prospecting tradition.

Most of the uranium finds were in the vast empty expanses that cover most of the Northern Territory, an area of prairie and desert inhabited mostly by aborigines, loners, eccentrics, rugged individualists, and idealists who want to help the native peoples. One loner of the Northern Territory was Jack White. In the late 1950s, he took up residence at an abandoned wartime airstrip 60 miles south of Darwin, at an unpopulated locality called Rum Jungle, which is located on maps only because it once contained a small copper mine. The cluster of mango and gum trees justifies its appelation as a jungle only by contrast with the barrenness of much of the surrounding countryside. The place acquired its name from an episode that occurred just after the copper mine closed down. The local store, the only one in hundreds of square miles, refused to give the miners credit any longer. One night, some of them broke in. They went after a keg of rum, and in their haste they smashed it, so that the rum all ran out. However, it flowed down into a clear water spring outside, and morning found the miners lying on their stomachs drinking happily from the rum-flavoured spring water. It was that kind of place.

White took out mining leases, thinking that he might dig out some copper from the old mine. Then he saw a pamphlet put out as an aid to uranium exploration, and realized that a picture of uranium ore was just like some samples of rock that he had saved. He sent the rock samples to the Bureau of Resources, and became the first man to receive the £50,000 reward the Bureau offered for finding uranium. Rum Jungle became a mining town again. Jack White remained in the outback.

The Rum Jungle find spurred a burst of exploration. Jack Gardner, a lifelong outback man, sometime prospector and resident of Darwin, had the idea that there might be gold at a spot he had walked over two years earlier, and when two friends suggested a weekend hunting trip, he persuaded them to come and look at it on the way. One of the friends had just bought a geiger counter, and he took it along, and they found, not gold, but a uranium deposit in sandstone that became the Adelaide River mine. They sold it to a mining company and became rich. A group of northerners formed the North Australia Uranium Corporation, raised money by selling shares, and sent exploration teams into the desert. Two Czech immigrants working for them (one of them a geologist who had worked at Joachimsthal) found the Upper Katherine Mine, 150 miles south of Darwin, happily just off the one highway that crosses the desert, and the company prospered. In another part of the country, two Queensland taxi drivers who tried their hand at weekend prospecting found a uranium deposit that one of them named after his wife, who had died of cancer two weeks before, and the Mary Kathleen mine became one of the richest in Australia. It was bought by an Australian subsidiary of the British mining company Rio Tinto Zinc.

Years later, uranium mining would become a moral issue to Australians. Some of them would look again at the uses to which their uranium was being put, and would worry about the implications of digging it out of the ground. Some would argue against doing so, and the issue would be debated fiercely between the main political parties and within the labour unions, and would be the subject of long official inquiries and demonstrations on the streets. But in those early, innocent days, in Australia and everywhere else, looking for uranium was simply a treasure hunt.

Chapter 3
ATOMS FOR PEACE

'Atoms for peace' sounds like what it was intended that it should sound like: a hopeful alternative to atoms for war. The potential for moving from the first to the second was underrated. This was a political mistake rather than a technical one, a matter of where the gaze was directed, and what things were excluded from the field of vision. 'Atoms for peace' was a slogan, a programme and a policy, and also a dream, that came at a time when the world was eager for just such a dream.

The programme was launched with excellent intentions, but sowed seeds that were to produce troubling fruits. It promoted an over-optimistic picture of the benefits of atomic power; it helped to stimulate an appetite for atomic power in countries which would not otherwise have wanted it so soon; and by training atomic scientists from many countries, it created influential groups of people with a vested interest in pressing their governments to build or buy atomic energy facilities, so that they could do the work that they were trained to do.

The enthusiastic effort to spread nuclear power for the economic benefit it brings marked a change of direction from the earlier policy of identifying atomic energy with atomic bombs, and guarding it tightly.

Inherent in the programme was the recognition that scientific secrets do not remain secret forever; that the most important secret of atomic energy was that it works, and this was revealed at Hiroshima. When, in later years, the programme came under increasing criticism, the Chairman of the AEC, Glenn Seaborg, said in a speech:

> Critics have overlooked the fact that it is impossible to keep science under lock and key for an extended period of time. . . . We realized that it might only be a matter of a few years before some other country or countries would be willing to provide nuclear materials and technology

> to others and to do so without firm assurances that such assistance would be used solely for peaceful purposes.

But this sounds like *ex post facto* justification. Statements of the time convey an impression, not of acceptance of the inevitable, but rather of enthusiasm for the spread of atomic power that sometimes bordered on the evangelical.

Some of the American scientists who promoted 'atoms for peace' were so uncritical in their approach that one can suspect that behind their zeal lay a sense of guilt at having created the atomic bomb, and a need to believe that, in unleashing atomic energy, they were Mankind's benefactors as well.

One member of the wartime Los Alamos atom bomb team said reflectively in a recent conversation: 'There was an element of guilt; anyway, for whatever reason, most of us believed that atomic power was the energy source of the future. We didn't focus on the problems. We simply didn't think the thing through.'

Another member of the Los Alamos team who voiced something of the same feeling is Frederic de Hoffman, a physicist who went on to start General Atomics, the first company anywhere dedicated to building atomic reactors. He wrote that the memory of making calculations for the dropping of the two atomic bombs on Japan made him 'determined to see that the peaceful uses of atomic energy would be fully exploited for the benefit of Mankind at the earliest possible time. Many of those who were with me at Los Alamos shared these feelings. . . . At General Atomics, for the better part of the two decades that followed, it was that very feeling that united us.'*

After the war, any modern-minded country had to be interested in atomic energy. Britain was a special case; the British Government felt that the United States had cheated it with the abrupt ending of the wartime partnership in atomic energy and withdrawal into secrecy, and it set about travelling alone the route they had

---

* He wrote this in a contribution to *All In Our Time*, a collection of reminiscences of the wartime bomb project edited by Ruth Adams. De Hoffman also wrote that the memory of making those calculations for Hiroshima and Nagasaki strengthened his resolve in helping to develop the much more powerful hydrogen bomb for America, which was not everyone's response to the experience.

previously covered together. Britain was still one of the big four powers, with global commitments, and hardly anyone doubted that if the atomic bomb existed, she should have it. A plutonium-producing reactor was built at Windscale on the north-west coast, and it went critical in 1950. The plutonium was used to make Britain's first atomic bomb, which was exploded on an island off the Australian coast in October 1952. But the heads of the programme were also thinking of atomic power, and they adapted the next plutonium-producing plant, at Calder Hall, next door to Windscale, so that it could also feed electricity into the national grid, the first full-scale nuclear power station anywhere to do so.

In America, for the five years after the war, the Atomic Energy Commission devoted almost all its energies to weapons research and production. But it also established a reactor division in 1949, and started work on four different experimental power reactors. It did studies on several uses for atomic power, including design work on an atomic aircraft engine. The one that took off was submarine propulsion; the Navy, in the person of Admiral Hyman Rickover, was enthusiastic, and a contract to build a submarine power reactor was given to Westinghouse.

France also started research on atomic power, though not yet on an atomic bomb. A Commissariat à l'Energie Atomique (CEA) was set up in 1945, and experimental atomic piles were built. In 1952, the Government announced a five-year programme for the construction of prototype power reactors, fuelled by natural uranium, like Britain's, because France did not yet have an enrichment plant.

But this was all going on inside research stations. For the world as a whole, atomic energy meant simply atomic bombs, and nothing else.

After the failure of the Acheson-Lilienthal plan and of any hopes of controlling atomic weapons internationally, new and sombre developments came in terrifyingly rapid succession: Russia's explosion of an atomic bomb in 1949; an atomic arms race between the two big powers; the Korean War, in which the two nuclear powers fought by proxy; and then the appearance of an entirely new kind of nuclear weapon, the hydrogen bomb, with a test explosion in the Pacific seven hundred times as powerful as the bomb that destroyed Hiroshima, an explosion that scattered radioactive fallout over hundreds of square miles and caused the death of a Japanese fisherman who was 80 miles away. This brought a new dimension to the nuclear peril. 'Atoms for peace' was a reaction to these developments.

A committee of scientists headed by Robert Oppenheimer submitted a secret report to President Eisenhower on the destructive effect of thermonuclear (*i.e.* hydrogen) weapons. The report recommended some new approach to Russia on preventing war, in view of the new potential for destruction, and recommended also that the public be informed of the dangers.

President Eisenhower accepted this, and he instructed C.D. Jackson, the former *Fortune* magazine editor who was his principal speech-writer and a key adviser on Cold War strategy, to produce a candid presidential message about the new nuclear weapons and their effects. Jackson formed an inter-departmental committee to work on what he christened Operation Candor. But the truth about thermonuclear weapons was horrifying (and it remains so). The committee turned out several drafts, but Eisenhower rejected each one as being too grim. After reading one draft, he said to Jackson: 'This leaves everybody dead on both sides with no hope anywhere. Can't we find some hope?'

It was Eisenhower himself who found the hope. During a holiday in Colorado, he conceived the idea of the two nuclear powers giving up some of their nuclear weapons material to an international body for peaceful use. He put this to Charles Cutler, an old banker friend and cabinet aide, and asked him to find out the technical problems involved. Cutler addressed a memorandum on the subject on White House stationery to Jackson and Lewis Strauss, the Chairman of the AEC. 'Suppose,' he began, 'the United States and the Soviets were each to turn over to the United Nations for peaceful uses X kilograms of fissionable material. . . .'

Jackson and Strauss set about working on this suggestion. Strauss proving more cautious and worried about giving any ground to the Russians. They took to discussing it over breakfast each morning at the Metropolitan Club, and Operation Candor became Operation Wheaties, so named informally after the cereal they used to order. As they worked out the plan, in consultation with Eisenhower every now and again, it was more than anything else an arms control measure. Strauss wrote in one memo that the problem was 'to find a formula for negotiation with Russia which would promote peace by partial or total disarmament.' For this was still the period of stark conflict, when there were no arms control agreements and no agreed common ground between the two sides.

The plan was predicated on the idea that uranium was and would remain in short supply, so that if America and Russia were both to hand over a sizeable quantity to be used for peaceful purposes only, this would limit the number of nuclear weapons they could build. Actually, the supply of uranium has never been a limiting factor in the nuclear weapons arsenals of the superpowers. But the expectation of a uranium shortage has persisted, and has influenced nuclear policy.

In all this, there was a strong element of Cold War competition with Russia for the goodwill of what today is called the Third World.

Eisenhower decided that the proposal should be launched with the widest possible fanfare. He told the Secretary-General of the United Nations, Dag Hammarskjold, that he would take up a longstanding invitation to address the UN General Assembly. The date was set for 3 December, the day on which he was to return from a meeting with the British and French prime ministers in Bermuda. Word was passed to the Soviet Government that this speech would be an important one, and its contents should be treated seriously. A draft was shown to Winston Churchill and the French Premier Joseph Laniel for their comments at the Bermuda meeting. The final version was typed on the plane flying back, and the aircraft circled New York for fifteen minutes while copies were mimeographed. Eisenhower drove straight from the airport to the UN building.

In a packed and attentive General Assembly, Eisenhower painted the still unfamiliar picture of the hydrogen bomb, and the perils of a world of two hostile powers in possession of it. He said there was a desperate need for some agreement to control the arms race. Then, in the last third of his speech, he launched into his proposal:

> It is not enough to take this weapon out of the hands of the soldiers. It must be put into the hands of those who will know how to strip it of its military casing and adapt it to the arts of peace. The United States knows that if the fearful trend of atomic military build-up can be

> reversed, this greatest of destructive forces can be developed into a great boon, for the benefit of all mankind.
>
> The United States knows that peaceful power from atomic energy is no dream of the future. That capability, already proved, is here today. Who can doubt that, if the entire body of the world's scientists and engineers had adequate amounts of fissionable material with which to test and develop their ideas, this capability would be rapidly transformed into universal, efficient and economic usage?
>
> I therefore make the following proposals: the Governments principally involved, to the extent permitted by elementary prudence, to begin now and continue to make joint contributions from their stockpiles of normal uranium and fissionable materials to an International Atomic Energy Agency. We would expect that such an agency would be set up under the aegis of the United Nations. . . . The United States would be more than willing – it would be proud – to take up with others principally involved in the development of plans whereby such peaceful uses of atomic energy must be expedited. Of those principally involved, the Soviet Union must, of course, be one.

The immediate reaction to this speech was tumultuous applause. Even the Soviet delegate joined in the clapping.

A few people, while approving the principle, worried a little about some of the details. In London, the Permanent Under-Secretary at the Foreign Office, Sir Harold (later Lord) Caccia, wondered about the strategic implications, and asked for someone to come from the Atomic Energy Research Establishment at Harwell to brief him on atomic energy. Two young physicists were dispatched to the Foreign Office, and they explained the principles of atomic energy to Caccia and a few of his colleagues.

As the meeting was about to break up, one of the two scientists asked the Foreign Office men a question about the Eisenhower proposal: 'How do you protect the fissile material?' The others asked what he meant by this, and he explained that Eisenhower had referred to handing over fissile material, by which he presumably meant uranium 235.

Why, the others wanted to know, should uranium 235 be protected? 'Because you can make bombs with it,' he explained. The Foreign Office sent a telegram to Washington supporting the atoms-for-peace idea, but suggesting that attention be paid to this problem.

The first Soviet response to the plan was negative. It came, after some preliminary sparring over procedure, in a 2,500-word reply to the American proposal sent in April 1954. Russia said the proposal evaded the key issue of banning nuclear weapons. This was the familiar Soviet ploy of rejecting a proposed better in favor of an unattainable best. Then the Soviet Government addressed itself to the problem of plutonium, though it did not mention the substance by name, and the exchange touched upon what has become the most sensitive question of nuclear technology. The Soviet Government evidently had also had a briefing. The Soviet note said:

> It is well known that it is feasible to carry out on an industrial scale a process of obtaining electrical power for peaceful needs by utilizing atomic materials, in which the quantity of the fissionable atomic materials applied in the process not only fails to decrease but, on the contrary, increases. And the harmless atomic materials are converted into explosive and fissionable materials which are the basis for the production of atomic and hydrogen weapons.

The United States, in its reply to Russia, raised the prospect of safeguards on nuclear power plants. 'The Soviet Union appears to assume,' the US note said, 'that any form of peaceful utilization of atomic energy must necessarily increase stocks of materials available for military purposes. In reality, however, ways can be devised to safeguard against diversion of materials from power producing reactors. And there are forms of peaceful utilization in which no question of weapons grade material arises.'

Eisenhower took steps to clear away domestic impediments to the carrying-out of his plan. He got Congress to amend the 1946 Atomic Energy Act so as to permit the transfer of fissile materials both to other countries and to private industry. Some Republicans opposed the plan because they did not want to give anything to the United Nations, but others supported it because they wanted American business to be able to exploit atomic energy internationally.

As things turned out, action did not wait upon Soviet agreement. Under-developed countries pressed at the United Nations for some follow-up to Eisenhower's bold words. The Taiwan delegate complained: 'I am afraid that a modest programme will not be able to sustain the universal enthusiasm and optimism that President Eisenhower's proposal has justly aroused.'

In November 1954, eleven months after Eisenhower's speech, the United States announced in the General Assembly that it was willing to give away 100 kilogrammes of fissile material for peaceful uses. However, the projected UN agency did not yet exist to receive it, so this material was to be given bilaterally to individual countries. This amount was increased massively in February 1957 to 20,000 kilogrammes, and later to 50,000 kilogrammes. The U-235 was not to be given in concentrated form, but in low enrichment uranium, never more than 20 per cent. Thus, sending 1,000 kilogrammes of U-235 would mean sending at least 5,000 kilogrammes of uranium, or more if it was enriched to only 3 per cent.

Meanwhile, talks were going ahead on the setting-up of the new international agency. The Soviet Union still said the renunciation of atomic weapons must be a precondition for any international plan for peaceful uses. Eisenhower announced that the new agency would be set up without Russia if necessary, though the original idea was scaled down. As UN Ambassador Henry Cabot Lodge told the General Assembly in November 1954: 'Originally, the United States visualized that the international agency would hold fissionable material itself. But after the Soviet rejection of the American proposals, all the negotiating states concluded that it might be preferable that the agency act as a clearing-house for requests made to it by various beneficiaries.'

When it became clear that an agency was going to be set up anyway, Russia switched its line and agreed to take part in the negotiations. The International Atomic Energy Agency came into being as a UN agency in July 1957. Its task, as stated in its charter, was to 'seek to accelerate and enlarge the contribution of atomic energy to peace, health and prosperity throughout the world,' while assuring that assistance provided through it was not used for military purposes. As things turned out, the IAEA did not take on the role of keeper of stocks of fissile materials, which was central to Eisenhower's original proposal. It was, however, to take on some of the task of ensuring that aid was not used for military purposes.

* * *

Britain was the first country to embark on a national nuclear power programme. Coal and oil were both more expensive in Britain than in the United States. Research into all aspects of atomic energy went ahead under the Ministry of Supply, and in 1954 the UK Atomic Energy Authority was established. The Secretary to the Cabinet, Sir Burke Trend, set up a working party of civil servants and they talked to scientists, and decided that this might be an important means of generating electric power. The result was the 1955 White Paper. This announced a plan to build nine atomic power plants with a total capacity of 2,000 megawatts. (A megawatt is a million watts; 1,000 megawatts is roughly the electricity requirement of a million people in Western Europe today.) As the men behind the plan saw it, this was only a first step, something to show for all the research and development that was being done. The power plants were all to be of the Calder Hall type, the Magnox. They would use natural uranium, since Britain had only a limited enrichment capacity, and would be cooled by CO_2.

As it turned out, the 1950s were a period of rapid development in electro-turbines. They were becoming bigger. In 1950, most were about 60 megawatts; by 1960, 500 MW turbines were being built. Since the nuclear power plants were to drive these new turbines, each reactor was different from the last. But the AEA was enthusiastic about the Magnox design, and felt that Britain was giving a lead to the world. In America, when the Joint Congressional Committee on Atomic Energy issued a report criticizing the Atomic Energy Commission for what it termed a 'sluggish' approach to nuclear power, it pointed to Britain as an example to be followed.

Within the AEC, the expectation was that nuclear power would become profitable in Britain and some other European countries before the United States because coal and oil were more expensive there.

In the wake of Eisenhower's UN speech, with its promise of 'universal, economic and efficient usage' of nuclear power, the AEC became more interested in it. It authorized Westinghouse to build a pressurized water reactor to provide electricity at Shippingport, Pennsylvania, based on the reactor it was building for the Navy's new submarines. President Eisenhower lent his prestige to the project by pressing a button himself to start construction, and he delivered a message saying: 'In thus advancing towards the economic production of electricity by atomic power, mankind comes closer to fulfilment of the ancient dream of a new and better Earth.' This was characteristic atoms-for-peace language. The US Post Office issued an atoms-for-peace stamp.

Now the public was treated to a stream of optimistic forecasts about the benefits that atomic power would bring. In Britain, the AEA said that atomic power would be producing electricity more cheaply than coal or oil by 1962, and from then on the price would fall steeply. The Chairman of the AEC, Lewis L. Strauss, forecast in a speech to science writers in 1954 that the day might come, thanks to nuclear power, when 'our children will enjoy electricity in their homes too cheap to meter'.

A book published in Britain in 1957, *The Economics of Atomic Energy*, by Mary Goldring, is interesting for the picture it gives, which was a common one at the time. At one point, it says: 'Atomic products are small in bulk; they do not need vast networks of trunk roads or ocean-going tankers to carry them. Their factories do not scar the countryside with railway tracks or darken the air with smoke.' It was a balanced and lucid book, but it contained not a single mention of nuclear waste, or the possibility of an accidental leak of radioactivity, and only a passing reference to the spread of nuclear weapons.

Others told how nuclear power would be science's latest gift to the ordinary citizen. The General Manager of the US Institute of Boiler and Radiator Manufacturers, Robert Parry, told a trade conference that miniature atomic reactors would be installed in homes as boilers within six years, and he passed around drawings. The little woman was not neglected. A leading American manufacturer of vacuum cleaners, Alex Lewyt of the Lewyt Corporation, forecast that atomic-powered vacuum cleaners would be on the market within ten years.

All this was just the supporting chorus to the international activities of the atoms-for-peace programme. The US Government gave grants to twenty-six other countries to help them build research reactors. Among the recipients were Brazil, Israel, Korea, Pakistan and Spain. It started selling the enriched uranium that it had set aside for this programme to other countries for use in research reactors or laboratory experiments – to India and Turkey in the early days. It also trained more than four thousand foreign scientists and technicians in America in nuclear science and engineering.

The high point of the international atoms-for-peace enthusiasm was the UN-sponsored Atoms-for-Peace Conference in Geneva in 1955, at which scientists from all over the world told each other what they knew about nuclear power. Those were the days of the Iron Curtain. Few Western scientists had ever met any Russians, scientific visits in either direction were very rare, and tourist visits to Russia were unheard of. For most Western scientists who attended the conference, talking with scientists from Russia and Eastern Europe was exciting, and even daring. The prospect of replacing secretive competition in atomic energy by open-hearted co-operation aroused great hopes, and scientists in several countries pressed their governments to de-classify material.

Just before the conference began, the Soviet Union scored some points in the peaceful atom game by suddenly inviting scientists from several countries to visit what it called the world's first operating nuclear power plant. What the visitors saw was hardly a power plant in the usual sense; it was a small 5-megawatt reactor at the Obninsk Atomic Energy Institute, and the electricity it produced must have cost a fortune. But it *was* feeding it into the grid, so, strictly speaking, Russia could claim a first in this area, and in the atmosphere of goodwill no one felt like cavilling.

Britain was looked up to at the conference as the country that was going to use nuclear energy on a national scale. The United States had the star exhibit at the conference: an atomic reactor built at the Oak Ridge National Laboratory, flown to Geneva in crates, reassembled there, and installed in a specially constructed building in the grounds of the Palais des Nations. It was the first reactor that many of the scientists at the conference had ever seen, let alone members of the public. The head of the delegation from West Germany, which had only just achieved sovereignty with the ending of the Allied occupation, was something of a celebrity: Otto Hahn, the man who, with Strassmann, had shown the world in 1938 that uranium atoms could be split.

More than one thousand papers were presented, and there was a welter of declassifying information, which one official present called a 'competitive strip-tease'. The most important information disclosed, at least as it seems in retrospect, was the Purex process of separating plutonium chemically from the other substances in used reactor fuel.

There was a lot of emphasis on the benefits that science could bring to the poorer areas of the world. The Chairman of the Conference was the Indian physicist Homi Babha. An Egyptian delegate said that atomic power could be used in his country to drill wells and raise water. A Pakistani said that in Pakistan it could drain marshes and power new industries. With this almost endless range of possibilities, it is small wonder that a Swiss woman went up to the US exhibit and asked where she could buy some uranium, because she had heard that it produced more heat than coal and she would like to put some in her boiler.

Chapter 4

THE NUCLEAR OPTIMISTS

The 1950s and '60s were still a part of the three centuries of the Enlightenment. The forebodings about rational Man and progress that had come already to a number of thinkers and artists and poets were not yet shared widely, though some recent events had planted doubts. There were few people around to challenge the view that science and its products would improve the human lot, or to suggest that they do not always work, or that their benefits may not always be real benefits, or that there may be a high price to pay for them. At the beginning of the 1960s, going to the moon still seemed to be not only a good idea, but also a natural one.

Governments began to take nuclear power seriously as an energy source worth substantial investment in the late 1950s. The message of 'atoms for peace' was being accepted in Europe more than in America, partly because Americans still saw their country as a land of abundance in natural resources, and there seemed to be no hurry about finding a new source of power. In Europe, the coal seams that had fuelled the Industrial Revolution were beginning to run down, and in fact, coal production in France, Germany and Belgium was to fall by half in the next twenty years. France followed its development work with a programme to build three electricity-producing reactors. The West German Government established the world's first Ministry of Atomic Energy, and subsidized research in industry and in universities.

The Soviet Union made a major commitment to nuclear power in 1957, with an ambitious programme of reactor-building. The Soviet Government foresaw no absolute shortage of fuel, but most of its coal and oil is located in Siberia and most of it is used in European Russia, so transport costs are high, and the nuclear alternative seemed a good idea.

In those days, a lot of the world's power stations burned oil. The Anglo-French-Israeli attack on Egypt in October 1956 closed the Suez Canal for several months; since most of Europe's oil was transported through this waterway, the closure brought shortages and rationing in several countries. This highlighted Western Europe's uncomfortable dependence on Middle East oil for power as well as for transport. Several countries looked again at possible alternatives for the production of electricity. The British Government responded to the consequences of its Suez war by tripling its nuclear power programme, setting a new goal of 5–6,000 megawatts of nuclear electricity by the end of 1956.

The six countries that were then joined together in the European Coal and Steel Community – France, West Germany, Italy, Belgium, Holland and Luxembourg – appointed a three-man commission to report on what atomic power could do for these countries. They produced an enthusiastic report, which envisioned the construction of nuclear power stations in the six countries to produce 15,000 megawatts of nuclear electricity by 1967. This would put the atomic power programmes of the six on a par with Britain's. Meanwhile, the foreign ministers of the six countries agreed in 1957 to set up a joint atomic power agency, to be christened Euratom. It came into existence at the same time as the EEC, on 1 January 1958, and was the first of the EEC's agencies. It was intended at first that Euratom should plan an integrated atomic power programme for the six, but European integration failed to live up to the hopes of those early days in this as in other areas, and it became only a safeguards and fuel supply agency. Today, in theory it purchases uranium raw and enriched, on behalf of member countries; in practice it normally just endorses agreements which member countries conclude.

In the United States, under the aegis of the AEC, there was still more research and development than production. At the AEC's test site in Idaho, scientists and engineers explored many possibilities of this new power source and fifty-one experimental reactors were built at the site. Work was carried forward on an atomic-powered aircraft, and a hangar for it was built in the desert. (This project was abandoned only in 1977; the problems of safety and adequate shielding for an atomic aircraft engine have never been solved.) The National Laboratory at Oak Ridge, Tennessee, where the first uranium enrichment plant was built in 1942, worked out plans for an agro-industrial complex, where industries would be sited in order to be near atomic power, and where, in addition, fertilizers would be manufactured, water desalinated, and food grown on previously fallow ground, all through the use of atomic power.

The AEC built a nuclear-powered merchant ship, the *Savannah*, and the Russians built a nuclear-powered ice breaker. But these were not followed by nuclear-powered fleets; they turned out to be much too expensive. For certain Navy ships such as aircraft carriers and submarines, it is a different story. The strategic advantage of being independent of outside fuel sources is worth the cost to the Navy.

There was even a programme to stage nuclear explosions for civil purposes, and this went on for years with test explosions underground. It was said that these could be used to release shale oil and natural gas deposits, among other things, and, if carried out above ground, work some massive feats of landscaping, such as diverting rivers. This programme also was dropped eventually.

Fission energy of a kind is used to provide power in small compact units, ranging from a space satellite to an Antarctic exploration station. But the only major use for nuclear energy so far is to produce heat which is used to boil water and drive a turbine to produce electricity. The nuclear power plant is hitched up to the electric generating system just like a power station that burns coal, oil or gas. From the steam stage onwards, it makes no difference what the power source is.

The consensus in America was still that atomic power would be useful first of all in other countries. In the wake of the Sputnik shock in 1957, when the supposedly backward Soviet Union beat the United States in putting a satellite into orbit, the Eisenhower Administration set up fifteen scientific advisory committees to deal with different subjects; not one of them dealt with atomic power.

American industrial firms were now starting to build reactors jointly with the AEC. They assumed that these would find markets abroad and that America would profit from other countries' experience in atomic power. Indeed, it was anticipated that this would be one of the benefits to America of the atoms-for-peace programme. An AEC report on this period said: 'It had been expected that the fruits of the US reactor development program might manifest themselves at an earlier stage overseas than they would in the United States. . . . Conversely, it has been recognized that a more rapid realization of economic nuclear power overseas will be of considerable value to this country.'*

A nuclear reactor is identified by one or more of the materials used in it. Thus, the first reactors to produce substantial amounts of power, the first generation of British reactors, are known as Magnox reactors, magnox being an alloy used as a cladding around the uranium fuel rods. The coolant in these reactors, the substance that takes the heat away from the core, is carbon dioxide, and they use natural uranium as fuel.

Canada entered the field with a reactor of its own design, for which was coined the vigorous and patriotic-sounding acronym CANDU, standing for Canadian Deuterium Uranium Reactor. Deuterium is heavy hydrogen, an isotope of hydrogen that has two electrons instead of one, and the reactor uses as a coolant heavy water, so-called because it is made up of oxygen and deuterium instead of ordinary hydrogen. Uranium comes into the name because the CANDU uses natural uranium as a fuel, (though recent studies have shown that it could use 1 per cent enriched uranium more efficiently).

Several different types of reactor were developed in the United States. The only one that came into wide use is the light water reactor or LWR. This uses ordinary water, which is called light water to distinguish it from the heavy water used in earlier, wartime reactors, both as a coolant and as a moderator. There are two sub-species of the LWR. One is the pressurized water reactor, or PWR; in this, the water is kept under pressure so that it can be hotter than boiling point and still not boil away, and so can carry away more heat. The other is the boiling water reactor, or BWR, in which the water is allowed to boil away so that the heat is transferred from the core by steam. The Westinghouse Corporation, which built the first PWR to power the submarine Nautilus, has concentrated on the PWR, and General Electric on the BWR. The PWR has become the bestseller worldwide.

One other reactor developed during this period deserves mention: the high temperature gas-cooled reactor, or HTGR. In this, helium is the coolant. The moderator is not put between the

---

* *Review of the International Atomic Policies and Programs of the United States: Report to the Joint Committee on Atomic Energy, 1960.*

uranium rods, as it is in other reactors, but is combined with the uranium in ceramic blocks than can withstand a much higher temperature than metals. The most significant feature of the HTGR is its fuel: 93 per cent enriched uranium. Two of these reactors have been completed in the United States, and one, at Fort St. Vrain, Colorado, is still operating; one was built in Britain, near Winreth, Dorset, as a multi-national European experimental project. Two small ones were built in West Germany, near Julich. It was realized after a time that the fact that this reactor uses bomb-grade enriched uranium as fuel outweighs any economic or technical considerations, all the more so as the United States sells fuel for it only sparingly, and will make no long-term commitment to supply highly enriched uranium.

The purchase of a reactor is the most expensive part of nuclear power. Compared with coal or oil-burning plants, a nuclear power plant has high capital costs and low operating costs. Furthermore, the location makes very little diffence to the operating cost; with coal and oil burning plants, the fuel is bulky and transporting it is costly, so that plants far from the coal fields or oil wells are substantially more expensive to operate than those closer to them. With a nuclear power plant as with any other kind of power plant, however, the location may make a difference to the cost of electricity to the consumer, since transmitting electricity over long distance is expensive. Reactors using natural uranium are usually more expensive to build than those using enriched uranium, but the fuel is cheaper.

These are all power reactors, but there are far more research reactors than power reactors, and more countries have them. More than one hundred have been sold by the United States on generous terms, under the atoms-for-peace programme. A research reactor is small, so that its power output is measured in thousands of watts rather than millions, or even less. The purpose of having one is to play with nuclear reactions: to find out what you can do with them, how to speed them up and slow them down, their effects on materials, and other things. Because one of these reactors is small, and because it needs as many nuclear reactions and as much neutron flow as possible in order that the effects may be analyzed, it uses very highly enriched uranium, usually 93 per cent enriched. When the United States exports one of these reactors, it agrees at the same time to provide a continual supply of bomb-grade uranium. It controls the use of this strictly.

* * *

Nuclear power is uranium power. Any country with a well-developed industry and technology can build a nuclear reactor. To make it work it must have uranium at least, and if it is an LWR it must be enriched. Most countries that wanted nuclear power at this time had to buy the uranium abroad, and they had to have it enriched abroad. An enrichment plant is difficult to build and expensive to operate. It can only be built economically on a very large scale. At this time and for many years afterwards, the United States was the only country with a plant that was enriching uranium for civilian use, so anyone buying an LWR was dependent on the United States for its fuel supply. The United States is still the world's leading supplier of enrichment services, and countries using LWRs still get some of their fuel from the United States. The AEC, which owned all the enrichment facilities in the United States co-operated with the reactor manufacturers by agreeing to enrich uranium for foreign buyers.

A key event in the success of the LWRs, and the extension of this dependence on America for fuel, was the association of the United States with Euratom's programme. It was American policy throughout the 1950s and '60s to encourage every move towards European integration, in order to strengthen the Atlantic Alliance. In the case of Euratom, the US Government agreed to finance the purchase by Euratom countries of American nuclear power plants totalling 10,000 megawatts capacity, through the Import-Export Bank. Most of these were to be built in Europe under license. The AEC played its part by signing a contract to supply enriched uranium for all these plants, and it signed it with Euratom, not with the individual countries. American reactors went out into the world with strong Government backing.

The assistance to the European programme provided valuable experience for American reactor manufacturers and power companies. The AEC report cited previously observed of the co-operation with Euratom: 'This program . . . provided an opportunity to construct large-scale demonstration atomic power plants in areas where fuel costs higher than those in the United States prevailed, thus prospectively diminishing operating costs.'

The first three power reactors that France built were gas-cooled and graphite moderated, like the British ones, and like those, they used natural uranium as fuel. The French Government wanted the other European countries to follow suit, rather than depend on the United States for reactors and also for a continuous supply of enriched uranium. But the Germans and others bought LWRS, encouraged by the AEC offer of fuel. Even Electricité de France, the national electric power company, used its freedom of action to join with a Belgian company in the construction of a US-designed PWR in Belgium that was to supply electricity to the national grids in France and Belgium.

On the Euratom Commission, the French representatives constantly questioned the wisdom of relying on the United States for enriched uranium, and wondered whether America might not apply constraints or conditions. But the German Fritz Efelfurt used to say, 'Don't worry. We'll buy nuclear fuel in just the same way that we buy coal or oil.'

A regular pattern of trade grew up: European countries and, later, Japan, would buy uranium in Canada, send it south of the border to be enriched to reactor grade in the AEC's plants, and then import the enriched uranium to use in their US-designed reactors. As nuclear power expanded, this flow of trade continued smooth and uninterrupted. The AEC felt secure in its monopoly of enrichment plant. One man connected with US Government policies in the 1950s and early '60s recalls today: 'The AEC could not conceive of the possibility that any other nation would be supplying U-235 fuel in substantial quantities.'

* * *

Nuclear power broke through into economic viability in 1963, or so it seemed. For the first time, a reactor was sold in America on a purely commercial basis. The Jersey Central Power and Light Company announced that it was going to buy a 515-megawatt BWR to be built by General Electric at Oyster Creek, New Jersey. GE guaranteed the price of the power station, subject only to correction for inflation, and the power company calculated that the Oyster Creek plant would produce electricity more cheaply than any other kind of power plant. There was to be no Government subsidy to either the manufacturer or the power company.

This announcement was hailed widely as the coming of age of nuclear power. From now on, it was said in the industry, nuclear power could stand on its own feet and look coal and oil in the eye even in the United States. Westinghouse jumped in with offers to sell PWRS at competitive prices. Both Westinghouse and General Electric were now selling nuclear power plants on what came to be called 'turnkey' terms, that is, the plant would be delivered to the customer just like a car, and all he had to do was turn the key, walk in and start it up.

In 1966, for the first time, more nuclear power plants were ordered in the United States than any other kind, and it was widely forecast that the day would come when nuclear power plants would be the only kind. The AEC now dropped all subsidies to nuclear power reactors (partly at the insistence of the American Coal Association). Things seemed to be moving in the same direction in Europe.

New figures were produced all the time that looked better and better, both for the cost of the plant and for its operation, and these promised cheaper electricity. There were expressions of triumph within the industry as new and better figures were achieved. So often were these figures for cheaper power repeated, and with such assurance, that it was almost forgotten that they were expectations rather than accomplished fact, expectations of what the cost of electricity would be when a still-unbuilt power plant was completed and operating several years hence.

Two authors who studied the nuclear power industry during this period describe the process: 'Government officials regularly cited the nuclear industry's analysis of light water plants as proof of the success of their own research and development policies. The industry in turn cited these same Government statements as official confirmation of its claims about the economic competitiveness of its product. The result was a circular flow of mutually reinforcing assertions that apparently intoxicated both parties and inhibited normal commercial skepticism about advertisements that purport to be analyses.'* The habit of living in the future tense was already ingrained in the nuclear power industry.

---

* from *Light Water* by Irvin Bupp and Jean-Claude Derian

The practice of treating prognostications as if they were measurable realities reached a kind of apotheosis in Britain in 1965. Sir Christopher Hinton, the towering, dynamic head of the UK Atomic Energy Authority's power division, wanted a new kind of nuclear power plant that produced higher temperature steam, so as to make the best use of the new and much larger turbo-generating plants which are powered by the steam. The Authority's engineers designed a new reactor to follow Magnox; like the Magnox, it was gas-cooled, but it used enriched uranium. It was called simply the advanced gas-cooled reactor, or AGR. Others in the AEA wanted Britain to turn to the American PWRS instead for the next series of nuclear power stations.

Discussion of reliability, safety and cost-effectiveness became enmeshed with national pride. (Arguments over the PWR persist to this day within the British nuclear power establishment, and arouse strong feelings.) The Government asked the Central Electricity Generating Board, as the principal user of nuclear power, to compare the two kinds of reactors. Cost figures for the AGR and the LWR were submitted by prospective manufacturers in Britain and by Westinghouse, calculated down to decimal places. They were analyzed as if they were auditors' figures, fixed immoveably in the past, instead of forecasts of the always indefinite future. The CEGB and the AEA put the British-built reactor ahead: electricity from the British AGR would be 10 per cent cheaper than from a PWR, they said. The Minister for Power, Fred Lee, announced this welcome news in Parliament, and, anticipating the export potential of the AGR, he said: 'We've hit the jackpot this time!'

The whole exercise was a technological fantasy. Nothing remotely like the degree of accuracy claimed can be attained in forecasts for machines not yet built. For all the sophisticated calculations, the comparison was between two biased guesses, both of which turned out to be wrong. Construction of the AGRS was started without a full-scale prototype. Every one had to be built on site, and every one presented new and daunting engineering problems. Five AGRS were ordered from three consortia. Cost overruns ranged between 50 and 500 per cent. The consortium that was to build one at a fixed price went bankrupt, and the CEGB had to take over the job. One AGR, Dungeness B on the Kent coast, lagged ten years behind schedule. On the ones that were completed, operating costs were higher than forecast, partly because reloading with fresh fuel rods without shutting down the plant, which was to effect big cost-saving, proved to be impossible, at least in the first years of operation. There have been no exports.

Everywhere the vague, rather starry-eyed optimism of the atoms-for-peace days was now translated into specific and technically-based forecasts, which turned out to be equally erroneous.

Engineers counted on a learning curve to reduce costs, as lessons learned expensively in earlier models are applied to later ones, a familiar phenomenon in many industries, but one that did not happen in nuclear power. They underestimated enormously the number of small engineering problems that would crop up, and the number of modifications that would be required in successive models.

James Stuart, the head of the firm which built some of Britain's reactors and a former Deputy Director of the Atomic Energy Authority, says now: 'Atomic power caught the imagination of the technologists in those days. Not the industrialists, who might have foreseen the industrial problems. The technologists were swept along with enthusiasm. The truth is, the possibility of a coal or oil shortage was a secondary consideration. It was brought in as a justification.'

The mistake was almost universal. In 1966, the Director-General of the International Atomic Energy Agency, Dr Sigvard Eklund, drawing together the information that individual countries supplied about their plans, reported that there would be 200,000 megawatts of nuclear-powered electricity in the world by 1980. The following year, 1967, he was pleased to be able to revise the figure upwards, to 300,000. By the time 1980 came, the actual amount was 130,000 megawatts.

Wrong assumptions were also made about alternative fuels, and this contributed to the errors about the economics of nuclear power. The nuclear power industry assumed that the cost of oil would rise during the 1960s, but in fact it fell. New oilfields were discovered in the Middle East, new drilling techniques proved to be more efficient, and the advent of the super-tanker lowered the cost of transport. More oil-burning power stations were built, and electricity generation from these became more efficient. If nuclear power was going to compete, then the price would have to come down even further.

The fact that nuclear fission produces plutonium and plutonium can be made into bombs was not forgotten. After all, the first big nuclear reactors were built during World War Two at Oak Ridge and Hanford specifically to create plutonium for bombs, and these were still operating for that purpose. So was the chemical separation plant at Hanford, at which the plutonium was extracted from the used fuel. It was still widely believed that plutonium created in the normal operation of a reactor, because it contains a high proportion of the isotope P-240, could not be used for an explosive. Nonetheless, the United States made it a condition when it enriched uranium for customers that it should not be separated chemically after use – or 'reprocessed', as this was coming to be called – without US permission. Subsequently, Canada made the same condition on the sale of its raw uranium.

Both countries waived this condition on the sale of fuel to Euratom, another example of preferential treatment for Euratom. The Euratom charter stipulates that nuclear materials that pass through it must be used only for peaceful purposes, and American and Canadian officials assumed that the separate members would watch each other like hawks to see that no one broke this rule: the French would certainly watch the Germans, and the Germans would watch the Italians.

Plutonium can also be used as reactor fuel, mixed with uranium. In the early 1960s, when the AEC enriched uranium, it bought back the used fuel rods and paid for the plutonium contained in them at 30 dollars a gramme. Later it stopped this practice, but the plutonium was still given a cash value as potential fuel, even when it was a part of the radioactive used fuel rods.

* * *

Of all the tricks that can be played with uranium, the breeder reactor is the neatest of all. It creates more nuclear fuel than it uses, so that it extends the use of a given quantity of uranium one hundred times. Most estimates say the world's known uranium resources are enough to last only for decades, but this would make them last almost indefinitely.

There has been so much discussion and argument about a

breeder reactor and its implications during these past few years that a layman might easily get the impression that it is an innovation. It is certainly new as part of a nuclear power programme: indeed, at this writing, it has not yet arrived, contrary to forecasts of two decades ago, which envisioned networks of breeder reactors in the main industrial countries by now. The first full-scale commercial breeder reactors are due to start operating in 1983, in France and the Soviet Union.

But the breeder reactor itself is almost as old as nuclear power. The first nuclear reactor to light a lamp was a breeder; at the Idaho test site, the Experimental Breeder Reactor, EBR-1 lit four 25-watt lamps in December 1951. Other experimental breeder reactors have been built, in America and other countries. The fact that they have only recently aroused anxiety, principally because of their association with plutonium, is one more instance of changing perceptions of an unchanging technology. We have not learned anything new about a breeder, except that it is more difficult to make it work than we thought, as with other reactors. Yet the things that we perceive today as dangers were overlooked in the past, perhaps because it seemed to be so triumphant a vindication of technology, solving the fuel problem for all time, and overcoming by ingenuity a fundamental limitation of the earth's resources.

A breeder reactor has a core made up of plutonium and a small amount of uranium, surrounded by a blanket of uranium. The uranium in the blanket can be ordinary uranium, or even depleted uranium from which most of the U-235 has already been removed, the leftover from enrichment. The neutrons radiating out of the core convert the uranium in the blanket into plutonium, and this can then be fed back into the reactor as fuel. It is plutonium that is bred in a breeder; the plutonium in the core creates more.

It is called a 'fast breeder'. This term, like 'enrichment', is misleading. A breeder reactor does not breed fuel fast. It takes about thirty years for one to produce as much plutonium as is fed into it. The word 'fast' does not apply to its breeding rate, but to a more arcane characteristic: nuclear fission is achieved with fast neutrons rather than neutrons which have been slowed down by a moderator, as they are in an ordinary, or thermal reactor. (It is called 'thermal' because the average energy, or temperature, of the neutrons after they have been slowed down is the average temperature of molecules in normal conditions. The energy content, or temperature, of fast neutrons is much higher.)

Structurally, the fast breeder is different from other reactors in that the core is much smaller and more concentrated, with no moderator. Since fast neutrons are less likely to split an atom than slow ones, and more likely to whizz right through it, more of them have to be put into the same space. The heat generated is more intense than in an ordinary reactor. Arnold Kramish, no critic of nuclear power but an analyst who has been concerned with it professionally for decades, called it 'almost a compromise between a thermal reactor and an atomic bomb'. Neither pressurized water nor gas can carry away the amount of heat that the core generates, so liquid metal is used, usually liquid sodium. This is difficult stuff to handle, and it is volatile. The energy density places greater strains on all the materials used. Also, if there were a major accident in a breeder reactor, the amount of radioactivity released would be much greater than in an ordinary reactor, though scientists working on a prototype breeder at the UK Atomic Energy Authority say such an accident is well-nigh impossible.

The fast breeder can only be a second generation reactor. There must first be thermal reactors to produce the plutonium to start it off, and it is more profitable if it uses depleted uranium from which the U-235 has already been extracted. Indeed, its ability to turn this otherwise useless material into plutonium is its great appeal. In Britain, the depleted uranium that could be used as fuel in fast breeders, and for no other purpose, has an energy equivalent of two hundred years' supply of coal.

In America, the programme has been dogged by accidents. EBR-1, at the Idaho test site, was put out of action by a core melt-down. Then the AEC and the Power Reactor and Development Company set out to build a prototype commercial fast breeder at Laguna Beach, Michigan, thirty miles south-west of Detroit. This was a historic milestone of another sort: it was the first time that major objections to a nuclear power plant on safety grounds were argued out at a national level. The AFL-CIO and four separate unions led the campaign to prevent the issuance of a construction licence, claiming that the fast breeder 'under present technological conditions is inherently unsafe'. A circuit court ruled in their favour, but the US Supreme Court reversed this finding, saying there was no reason to assume that safety measures were not effective. The reactor was christened Enrico Fermi 1, and began operating in August 1963. It gave trouble from the start, and was finally wrecked by a partial core melt-down in October 1966. This could have had serious consequences – just how serious is a matter of argument still. Some people reported that when the melt-down occurred, police were alerted to be ready to evacuate Detroit, though the authorities say that no such message was ever sent. The episode was the subject of a book with the sensational title *We Almost Lost Detroit*, which is a weapon in the anti-nuclear armory.

No other country has had quite the same amount of trouble, but progress has been slow. Britain built a research breeder reactor at Dounreay, a remote site on the North coast of Scotland, back in 1955; its container is a white sphere, so striking visually against the bleak shoreline that it has become almost a symbol of Britain's nuclear power programme. Then more recently, it has built a larger prototype on the same site. France and the Soviet Union have prototypes operating – the French one produces 230 MW and is called Phénix, the predecessor of the Super-Phénix which is due to start producing electricity in 1983. In the United States, the AEC began building a fast breeder reactor at Clinch River, Tennessee, near Oak Ridge. This project was cancelled by President Carter and revived under President Reagan.

As the costs of reprocessing used fuel to extract the plutonium, as well as the cost of breeder reactors themselves, have risen, the economics of the breeder have become less certain. Some now say that it will save uranium at a much higher price than the price of the uranium it saves. It was an extreme view, but by no means a unique one, when the British physicist Lord Bowden said in a brief discussion of the subject in the House of Lords in July 1980, that the fast breeder would prove to be, 'like Concorde, a technical triumph but an economic disaster'.

However, the several governments that are pressing ahead with plans for a breeder reactor programme are not motivated only by economic considerations, nor was President Carter when he cancelled the Clinch River project. For many countries, a principal attraction of the breeder is that it would make them independent of any outside source of fuel, and also unworried about any scarcity that may come. The US Government's renunciation of breeder construction was part of a campaign to dissuade other countries from building breeders now because they involve the dissemination and stockpiling of plutonium. The clash between these two aims dominated much of the inter-governmental discussion of nuclear power for years, and the issue is still a live one today.

Chapter 5

THE TREATY BRAKE

The principal internationally-agreed instrument for checking the spread of nuclear weapons is the Nuclear Non-Proliferation Treaty, the NPT. It is a discriminatory treaty; it could hardly not be, since the very term 'proliferation' is discriminatory, implying as it does that the development of nuclear weapons up to a certain point is acceptable but its spread to other countries is a peril to be avoided. To the United States, the proliferation problem began when Russia developed the bomb; to the superpowers, it began when France and China developed theirs; to other countries, it began with the first atomic bomb test at Alamogordo, in July 1945. One country's self-defence is another's proliferation. The treaty discriminates in its application between those countries that have nuclear weapons and those that do not. Unlike most treaties between haves and have-nots, its declared aim is not to bridge the gulf but to perpetuate it. Considering this, it is perhaps remarkable that any of the have-nots signed the treaty, and is a testimony to their desire for a stable world.

The essence of the NPT is simple: nuclear weapons states promise not to give nuclear weapons to anyone else, and non-nuclear weapons states promise not to acquire them. Its effects are limited because, for the most part, the signatories are promising not to do what they had no intention of doing anyway, or are incapable of doing. The United States, Russia and Britain, the three nuclear weapons states which have signed the treaty, would not pass on nuclear weapons to any other country with or without the treaty. Before the treaty was signed, the Soviet Union had already made the biggest sacrifice that any country has made in the cause of non-proliferation: it created a rift with China, then its great Asian ally, rather than help China build its own atomic bombs. As for the non-nuclear weapons signatories, most do not have the capability to build nuclear bombs anyway. The world is not much changed because Gambia, El Salvador and the Vatican City, all signatories of the treaty, have promised not to build nuclear weapons.

What counts, and what has always counted, is the signature of countries that can build nuclear weapons. Among these, Australia, Canada, East and West Germany, Japan and Sweden have signed. This ratifies their previous decisions not to build nuclear weapons, but it gives this decision the underpinning of an international commitment, and creates one more threshold to cross before the policy is changed. Argentina, India, Israel, Pakistan and South Africa are potential nuclear weapons states that have not signed the treaty.

France and China, which have not taken part in any formal arms control agreements, have not signed the NPT either. Both say in different ways that its discriminatory character is unacceptable. France has said it will nonetheless abide by the terms of the treaty. However, it offended against its spirit for some years by supplying nuclear materials to Israel on terms which were kept secret. China, on the other hand, said that it did not accept the validity even of the goals of the treaty, and there was nothing wrong with other states acquiring nuclear weapons providing that they were peace-loving states. But, when the Libyan President Gadaffi visited Peking, and asked Chou En-Lai, with extraordinary naïveté, whether Libya could buy an atomic bomb from China, Chou declined to sell him one, and said that every country must take care of its own defence.* And China has never taken any step towards helping another country acquire an atomic bomb. One might say that France has always supported the NPT in principle but not in practice, while China has supported it in practice but not in principle.

The Nth country problem – the problem of what to do when nuclear weapons spread beyond the big powers to a sixth, seventh and Nth country – was already a popular subject of discussion among academic strategists in the late 1950s. Several academic institutions produced papers on prospects of the nuclear spread; most predicted a more rapid spread of nuclear weapons than in fact has occurred. A typical forecast was the report by the US National Planning Association in May 1958 that there would be between eight and twelve nuclear weapons states by 1970.

The first non-proliferation treaty was the test-ban treaty signed

* This was stated by the highly-placed Egyptian editor Mohamed Heikal in his book *The Road to Ramadan*, and it has not been denied by either party.

in 1963. Of course, non-proliferation was not its only purpose, nor even its principal one, but since testing was regarded then as an essential step in producing nuclear weapons, it was assumed that a nation that signed could not build nuclear weapons. Certainly China, which had not yet tested its first bomb, saw it this way. The Chinese Foreign Minister Chen Yi described the treaty as 'a plot to prevent China from acquiring her own means of self-defence'. The plotters, as China saw it, inevitably, were the United States and Russia, and signature of the treaty widened the Sino-Soviet rift.

Proposals on the non-dissemination of nuclear weapons, which was the term then in use, were introduced in the UN General Assembly in the late 1950s and early '60s and it may seem surprising in retrospect that these came first of all from Third World countries, the have-nots in nuclear weapons terms. India, which later refused to sign the NPT, was one of the first to urge non-dissemination measures on the two superpowers. These finally produced proposals of their own in 1965, and after some hard negotiation, they agreed on the NPT in 1968.

The hard negotiating did not concentrate on the points that today seem important, as is so often the case in this field. For one thing, anxiety about the spread of nuclear weapons was focussed principally on West Germany, and not on the more developed of the Third World countries, as it is today. For another, the discussion was mostly not about possible pathways from civil to military nuclear power, but about the transfer of nuclear weapons from one country to another. The principal subject of discussion was the increasingly complicated and increasingly unrealistic proposals for a NATO nuclear force that the United States was putting forward to its European allies. All involved an American veto over the use of these weapons, but nonetheless the Russians were worried that any sharing agreement would bring Western European fingers, and in particular West German fingers, closer to a nuclear trigger. They examined every draft of a non-proliferation treaty carefully to check for loopholes through which this eventuality could slither.

Their worries were misconceived. The aim of American proposals for a NATO nuclear force was the same as the aim of the projected treaty: preventing the spread of nuclear weapons. America had failed to dissuade France from building its own nuclear weapons, and assumed, wrongly, as it turned out, that other Western European countries might want to follow the same path. It was trying to head off any such ambitions by offering instead participation in a multi-national nuclear force over which it would have the veto, the shadow of nuclear weapons power in place of the substance. But there was not sufficient interest among the European allies, and the NATO force plan never took off.

Most of the treaty consists of measures to ensure the carrying out of its terms, and of measures designed to mitigate, or give the appearance of mitigating, its discriminatory aspects. Under the treaty, all nuclear facilities acquired by a signatory, or transferred by a signatory, must be inspected to ensure that they are not used to make atomic bombs. Responsibility for inspection was entrusted to the IAEA which had already built up a corps of inspectors and a fund of inspection techniques. Facilities in the nuclear weapons states are not subject to inspection. V.C Trivedi, the Indian disarmament specialist who led India's criticisms of the treaty, said in a much-quoted official statement: 'The institution of international controls on peaceful reactors and power stations is like an attempt to maintain law and order in a society by placing all its law-abiding citizens in custody while leaving its law-breaking elements free to roam the streets.' It might be a more apt metaphor to say that it is like innoculating people against a disease but not bothering to innoculate those who already have it. The object of inspection is to ensure that a country does not use civil nuclear facilities to develop a nuclear weapon clandestinely. Countries that already have nuclear weapons and acknowledge the fact would have no purpose in doing so.

International inspection is a certain amount of trouble for the plant concerned, and causes some anxieties about commercial secrets. Because of this, and because this is one more discriminatory aspect of the treaty and so an extra irritant, Britain and the United States have now agreed to allow inspection of their non-military facilities, like a parent who swallows a dose of nasty-tasting medicine that he does not need in order to persuade a child to take some.

Article VI contains a promise by the nuclear weapons states to narrow the gap in military terms between the haves and have-nots, by negotiating 'in good faith on effective measures relating to cessation of the nuclear arms race at an early date and to nuclear disarmament'. By the time the treaty was drawn up, the terms 'horizontal proliferation' and 'vertical proliferation' had come into use, the first meaning the spread of nuclear weapons to other countries, the second the piling up of arsenals of nuclear weapons in countries that already have them. The non-nuclears insisted that if they were to forego horizontal proliferation, the nuclear powers must forego vertical proliferation.

The non-nuclears feel that even in its own general terms, the promise of article VI has not been kept, since all the nuclear weapons powers have increased their arsenals. And it is difficult to see this promise by the nuclear weapons states as anything more than an empty gesture. When they negotiate over arms control, this promise is not an input (though President Carter listed this as one factor when he asked the Senate to ratify SALT 2). When, at the five-yearly review conferences, the non-nuclears attack the nuclear weapons states on this score, the hapless diplomats of the former can only point to the few arms control agreements that have actually been negotiated, such as SALT I and the threshold test-ban treaty. In fact, when two sides have yet to reach an agreement on something, there are no demands that can logically be made to both of them. Faced simply with the injunction to 'negotiate in good faith', each can only say, 'Well *I'm* doing so'.

In any case, the link between vertical and horizontal proliferation is tenuous, if it actually exists. A nuclear build-up by the superpowers may affect the psychological atmosphere in which decisions on nuclear weapons are taken, but it is difficult to imagine a country taking a decision to build a nuclear bomb because the United States has eight thousand nuclear warheads instead of six thousand.

Much time has also been spent on Article IV of the treaty, and arguments over this are more substantive. This clause seeks to assure that non-nuclear weapons countries, in accepting an inferior status militarily, will not be accepting an inferior status industrially as well. It asserts the 'inalienable right' of all the parties to develop nuclear power for peaceful purposes. It goes further, and pledges all the parties to facilitate 'the fullest possible exchange of equipment, materials and scientific and technological information for the peaceful uses of atomic energy'. New restrictions which countries place on the export of materials or technology because they could pave the way to nuclear weapons production sometimes give rise to objections on the grounds that they breach article IV.

Two major international inquiries have used this article as a reason for urging the expansion of nuclear facilities. In Britain, the Parker Inquiry on the expansion of the Windscale plant to reprocess used fuel from abroad concluded that the NPT placed an obligation on Britain not to restrict such services. In Australia, the Ranger Inquiry set up to determine whether Australian uranium should be mined and exported reached a similar conclusion.

There is also a clause in the treaty about peaceful nuclear explosions, since there was still a thought in people's minds that these might be useful. There was some arguing during the negotiating stage about whether a non-nuclear weapons state could stage a peaceful nuclear explosion under the terms of the treaty. But a nuclear explosion is a nuclear explosion, and whether it is a weapon or not depends solely on the use to which it is put, so the treaty as it is finally written bans all 'nuclear explosive devices'. However, as a concession to the view that these could be valuable one day, it pledges the nuclear weapons states to make these available to non-weapons states for peaceful purposes at cost. But though this issue has been raised from time to time, no non-weapons state has ever made a specific request for a peaceful nuclear explosive, and the issue is vanishing.

The NPT was signed in London, Washington and Moscow simultaneously on 1 July 1968. It came into force in 1970, when forty countries had signed and ratified it. Under its terms, a review conference is to be held every five years. Starting with the first, in 1975, the pattern emerged of nuclear weapons haves versus have-nots. One debate after another saw the superpowers standing shoulder to shoulder in fending off the others' demands.

A group of non-nuclear weapons states made a suggestion to reduce any security disadvantage they might suffer by renouncing nuclear weapons. They asked the three nuclear powers to promise not to use nuclear weapons against any non-nuclear country that had signed the NPT, except in the case of a general nuclear war in which both were involved. This proposal was backed, not only by Third World countries, but by some which normally line up with the United States, such as Australia and Mexico. But the three nuclear powers rejected it. The American reasoning, expressed only in private, was that in the wake of the Vietnam débâcle (this was in 1975) nothing should be done which could weaken even theoretically the strength of an American guarantee to its allies – to Japan in the event of an attack by North Korea, for instance.

America made a slight gesture towards the sentiments behind this request in 1978. It promised that it would not use nuclear weapons against a non-nuclear NPT signatory unless that country attacked the United States or one of its allies and was allied to a nuclear weapons state. The exception would include just about any serious shooting situation in which the United States could get involved. One Administration official, according to the *Washington Post*, asked what the effect of this would be, said, 'It means that if a Ruritanian official kicks a GI, we can't use nuclear weapons in retaliation.'

The NPT is important principally because it establishes a norm of international behaviour: the non-acquisition of nuclear weapons, and the export of nuclear materials only under safeguards. Most nuclear material and equipment is now sold abroad only under safeguards, so that even countries that have not signed the NPT have most of their nuclear facilities under safeguards because they could not get them any other way.

The treaty has a clause permitting any signatory to withdraw at three months' notice if it feels its national security is in danger. But at least a government must take a decision to do this; its members must debate the action, in public or in private, and in any weighing up of the pros and cons, the affront to the other treaty signatories is a con that must go on the scales. Participation in the treaty can serve as a brake on any move in the direction of nuclear weapons.

Most countries that export nuclear materials now insist on bilateral agreements on safeguards and inspection as a backstop to the NPT. Thus, if the receiving country were to leave the NPT, after giving the required three months' notice, it would still be bound by its agreement with the supplier country to accept international inspection of the facilities.

Of course, a country could still quit the treaty and abrogate any bilateral agreements on inspection unilaterally. Then only military force could reverse the situation. So long as nations are sovereign, the nightmare possibility behind all talk of inspection and safeguards over nuclear facilities that are sold is that one day the IAEA inspectors may turn up and find a nuclear power plant surrounded by troops, and be turned away at the gate at bayonet point.

One can construct a number of scenarios in which this kind of thing could happen. For instance, the Shah of Iran had an over-ambitious programme for nuclear power development. The revolutionary régime that overthrew him scrapped it. But suppose the revolution had occurred a few years later. Nuclear power plants and, more important, a reprocessing plant would have been in operation in Iran. They would be sealed and visited regularly by IAEA inspectors, probably from Western Europe because that is where most inspectors come from. Surely the revolutionary authorities, fired as they were by Islamic zeal and anti-Western nationalism, would not hesitate to bar the way to Europeans who would be acting as governesses to their infant industries, and to denounce the treaties which gave them the right to do so as 'nuclear neo-colonialism'. Other countries besides Iran may and probably will undergo nationalist and anti-Western revolutions.

One can imagine other situations in which a country would leave the NPT. A political party might arise with an aggressively nationalistic programme, proposing measures which affect international ties and institutions, as Gaullism did in France in the late 1950s. In slightly changed circumstances, one can imagine such a party arising in Japan, for instance, or Greece. Among other things, it would denounce the NPT for the subservient status it imposes on the country. If this party came to power, it might then announce withdrawal from the NPT, under the terms of the treaty itself, all the while disclaiming any intention of building nuclear weapons. This might simply be a gesture to assert national pride, and a new status in the world, or else the first step in a military programme, or, like Hitler's denunciation of the Versailles Treaty and re-occupation of the Rhineland, it could be both.

* * *

Inspection of nuclear facilities is so much an accepted part of the nuclear power scene today that we can easily forget how sharp a break it is with the conventions of national sovereignty. There is no other circumstance in which foreigners have the right to go into a country and check on an industrial plant. Yet in most countries that have any kind of nuclear equipment, regular visits by the man from the agency, carrying his radiation counter like a briefcase, are a normal part of the operation.

The first formal safeguards arrangement on a nuclear reactor was a part of the first United States atoms-for-peace agreement, signed with Turkey in June 1955. The United States provided a research reactor plus fuel to Turkey on favourable financial terms, and the Turkish Government promised not to use the material for military purposes. To ensure that this promise was kept, the Turkish Government agreed to allow AEC inspectors 'to observe from time to time the condition and use of any leased material, and to observe the performance of the reactor in which the material is used'.

These days, observation is much more sophisticated than it was then, and inspection arrangements more detailed. Much of it is done through electronic devices, by remote control. In those days, inspectors were prepared to allow for a 2 per cent error, which means that up to 2 per cent of the fissile material could be diverted without anyone being certain that this had happened. Nowadays, they are ready to say that they are certain of 99 per cent of the material, and 100 per cent in most cases.

National governments still have trained inspectors. For one thing, India, until recently, accepted American and Canadian inspectors but not those from the IAEA. Euratom carries out inspection in its area by agreement with the IAEA, and has its own inspectors' corps. But the IAEA is still the principal inspection agency and sets standards.

IAEA inspectors are nuclear engineers, who join the agency on a two-year contract. Most stay more than two years, but all the same, there is a brisk turnover. The inspector is assigned to one country or region, so that he will set out from Vienna at regular intervals for the same place, Iberia, or Japan, for instance. No inspector is assigned to his own country. In some cases, where a nuclear facility is particularly sensitive, an inspector is installed permanently at the site, and takes up residence nearby. The policing is routine, and usually cordial, and much of it is done by machines, but it is policing just the same.

Inspection is deterrence, not defence. It is not designed to prevent cheating, but to ensure that it can be detected once it has taken place. A key phrase in the statement of IAEA safeguards objectives is 'timely warning'; detection should be well in advance of the time that it takes a country that has diverted some material to put it to malign use.

What the inspector is checking for is diversion. He is there to make sure that all the fissile uranium or plutonium that appears in any part of the system is used for its stated purpose, and none is removed surreptitiously to be used for something else. This means checking on the fissile material at every stage of the system.

If a country has a fuel fabrication plant, then this must be inspected. An inspector checks on the uranium ore concentrate, yellowcake, when it arrives from the mill. He must work out the proportion of uranium, which will be between 40 and 80 per cent, using a scintillator to measure the radiation level and calculating the amount of uranium from this. If there is an enrichment plant, an inspector checks on the amount of uranium going in and out of it. When it is made into fuel rods, an inspector checks on the amount of uranium in the rods, and then makes sure that all the rods are fed into reactors.

When the fuel emerges from the reactor, he checks on its storage. If the fuel rods go to a reprocessing plant, then an inspector checks on this operation also. This is the most demanding task of all, because what goes in at one end is highly radioactive and cannot be measured directly, and what comes out at the other end is a number of streams of liquid which are radioactive in different ways. Chemical samples are taken of the solution at several stages of the process.

A sample is even taken of the liquid with which the plant is washed out, to see whether any of the material is left clinging to the sides of tanks or tubes. An IAEA manual observes: 'Scope for a would-be diverter lies in inadequately performing these wash-out procedures so that all material held up in the plant does not accumulate in the designated measurement points. Careful verification of these procedures is essential.' There is nothing perfunctory about these inspections.

A key device in inspection is the seal. Once an inspector has checked on a certain part of a plant, he seals it off, so that no material can be taken out before his next visit. The seal is an electronic device; any attempt to open it will be immediately detectable. The IAEA is using more and more remote surveillance devices. The principal one is a movie camera trained on an area in which the fuel rods are stored, connected to a photo-electric cell and activated by any movement across its line of vision.

At the top policy-making level, most IAEA officials say that the problem of safeguards is primarily political rather than technical, that it is a matter of getting countries to accept safeguards. They feel that a country would not accept safeguards if it intended to cheat on them, because it would be very embarrassing if it was caught cheating. However, one cannot assume safely that cheating will not be a problem, and that it will never be tried.

There are conceivable circumstances in which a government might want to divert some fissile material clandestinely in order to create a little stockpile to make weapons, particularly if it could blame some group of people in the nuclear power industry or its military establishment if it were caught. Indeed, one of these groups might decide by itself to try to divert some nuclear material, perhaps in a country that has not signed the NPT but has had to accept inspection in order to acquire the beginnings of nuclear power.

Certainly the members of the IAEA's inspection division put all their ingenuity into their work as if cheating were a constant threat, like those technicians of nuclear deterrence who exert great efforts to meet a surprise missile attack, while somewhere else politicians and strategists agree on the unlikelihood of its ever happening. The inspection division of the IAEA has a Systems and Plans Section, in which people think up new challenges for the safeguards system.

For instance, some of these pointed out that, whereas it is always possible to tell whether a seal has been broken, someone fiendishly clever could break the seal, remove it altogether, make an exact duplicate, and put the virgin duplicate model in its place. So now seals have fibre-optical loops which give off a unique pattern of light so that an inspector can check instantly on whether the seal he is looking at is the one he implanted. A newer device still is a seal which has a display of figures changing constantly according to a coded pattern that only the inspector knows, so that any substitute would not be displaying the correct figures.

The Agency has been situated in Vienna since its beginnings, and at the end of 1979, its headquarters were moved to the new International Centre among the suburbs on the left bank of the Danube, a set of office buildings arranged in concave and convex shapes, so that the hiker down its bureaucratic corridors finds himself leaning first to one side and then the other, like a sailor on a swaying ship. The agency's staff of 1,500 share the buildings with other, smaller UN agencies. Like the staffs of other international organizations, these come from East and West, both politically and geographically, but here have the same technical and bureaucratic concerns and a similar lifestyle.

* * *

The IAEA has become identified widely with the NPT and safeguards, but policing nuclear power is not its only function, nor even, according to its statute, the principal one.

Its statute charges it with twin tasks: 'to seek to accelerate and enlarge the contribution of atomic energy to peace, health and prosperity throughout the world'; and also to 'ensure, so far as it is able, that assistance provided by it or at its request or under its supervision or control is not used in such a way as to further any military purpose.' These two aims are not strictly incompatible, but they are divergent, so that pursuing them both can sometimes create a strain. It is rather like a liquor dealer becoming chairman of the local branch of Alcoholics Anonymous. The liquor dealer, assuming he is a reasonably good citizen, does not want to create alcoholics in the course of his trade, and it is not the goal of Alcoholics Anonymous to stamp out the liquor trade, but adopting both roles can be difficult, and lead to the suspicion by interested parties that one is being emphasized at the expense of the other.

The Indian V.C. Trivedi warned specifically on one occasion, in a speech: 'The agency must not let its control functions predominate over its promotional ones.' And there are frequent arguments within the agency's thirty-four-member board of governors about the budget, in which representatives of the underdeveloped countries argue that funds should not be taken from the Technical Assistance Division to pay the overheads of the Inspection Division.

It certainly takes its promotional functions seriously, and technical assistance also. At any given time, there are between thirty-five and forty IAEA experts in the field who have been sent by the IAEA to give advice and assistance. At one time recently, which was quite typical, these included a Dutchman advising Morocco on the training of nuclear physicists; a British specialist in reactor engineering in Spain; a Polish specialist in medical radio-isotopes in Jordan; an Australian advising the Ecuadorian Government on uranium exploration; and an Indian specialist in radiation-measuring instruments in Indonesia.

The IAEA serves a useful function as a clearing-house for ideas and information. It issues guidelines in areas where new concern arises, such as the transport of nuclear waste, and the protection of nuclear materials against theft. It holds conferences in every part of the world on many different aspects of nuclear energy. Third World member countries are constantly pressing it to devote more attention to uses of nuclear energy other than power production, in which most of them have little interest, such as using radiation to improve agriculture, and in medicine; many of its conferences and papers are devoted to topics such as these. Its literature and its officials always promote the idea that nuclear energy and nuclear power are a good thing.

There is another non-proliferation treaty in existence, the Treaty of Tlateloco, which is designed to bar nuclear weapons from Latin America. It was actually drawn up before the NPT, in 1967. It is named after the suburb of Mexico City in which it was drafted. It is not a self-denying edict on the part of Latin American countries. The idea is that it serves as a guarantee for each against another acquiring nuclear weapons.

It suffers from the same weakness as the NPT, only more so. The two major powers in the continent, the only two that could conceivably construct nuclear weapons in the near future, Argentina and Brazil, are not bound by it. Argentina has signed but not ratified. Brazil has signed and ratified but unlike most of the other signatories, it has not waived a requirement of the treaty that it does not enter into force until all the countries of the region have signed it. Since Cuba has not signed it (and says it will not while the United States retains the base at Guantanamo) it has not entered into force yet so far as Brazil is concerned.

For a while, it looked as if the only country to accept meaningful restrictions under the treaty would be the United States. Countries which have territories in the region were asked to sign a protocol promising not to introduce nuclear weapons into these territories, and Britain, France and Holland did so. The United States signed in 1977, which would mean barring American nuclear weapons from American bases in the Caribbean, including Guantanamo and Puerto Rico. But the treaty has not yet come before the Senate for ratification.

Chapter 6

BENDING THE RULES

On the morning of Saturday, 14 May 1974, the Indian Foreign Ministry telephoned the embassies of all the major powers and said it wanted to see their representatives immediately to deliver a message. The first man to go to the Foreign Ministry in response to this telephone call was the Soviet ambassador. The second appointment was made with the Canadian chargé d'affaires, William Jenkins, who was temporarily running the Canadian High Commission in New Delhi. Setting out, Jenkins wondered what this Saturday morning call could be about. His guess was that the Indians were going to say that they were withdrawing from the International Truce Commission in Vietnam, on which the Canadians were also participants, because the commission could not function properly.

As he drove up the driveway of the Foreign Ministry building, he saw the Soviet ambassador leaving and noted a shocked look on his face. Then it occurred to him that the call might be about something more important. A few minutes later, the director of the Foreign Ministry was telling him that a nuclear explosion had been carried out at eight o'clock that morning in the Rajasthan Desert. The director said the explosive device was not a weapon, but was intended for peaceful purposes. Jenkins asked whether there was any more information that he could have, and he was told that there was not.

He went straight back to the High Commission and cabled the news to Ottawa, where it was received with shock and outrage. He was instructed to ask the Indians whether the explosive material was extracted from used fuel from a reactor supplied by Canada. Several weeks went by, and then he was told that it was. Even before this reply was received, Lorne Grey, the Chairman of the Government-run Atomic Energy of Canada Ltd., a one-time engineer who believed strongly in nuclear power for peace, called on the Indian High Commissioner in Ottawa, and told him heatedly that the Rajasthan explosion was 'an act of betrayal'. The High Commissioner said he could not accept this language.

The Indian explosion produced enormous reverberations internationally, which are continuing still, and started a re-evaluation of the relationship of nuclear power to nuclear explosives. To see why it produced such an effect, and why it took place at all, one must go back in time to look at India's attitude to nuclear power, and other countries' attitudes towards India.

Although India is one of the poorest countries in the world and the second most populous, it has always had a Western-educated élite and a middle class from which this could draw. It was able to create and sustain a system of democratic government more successfully than most other newly independent countries, and able also to develop high technology industries which are a feature of the Indian economy today, and which distinguish it from most other countries in the same economic bracket.

Pandit Nehru was a characteristic member of this élite. Another was Homi Babha, a physicist of international reputation, who virtually created singlehanded India's nuclear power programme. A plump, moon-faced man of medium height and serious demeanour, Babha was a Parsee, and the son of a wealthy family. At Cambridge University in the 1930s, he made important contributions to quantum theory and to our understanding of cosmic radiation. He went back to India in 1940, and with money from the Tata industrial empire, he established the Tata Institute of Fundamental Research. It was staffed by Western-trained scientists and he himself made sure that young Indian scientists were trained there, and he became the acknowledged mentor of a generation of Indian physicists. In 1944, before the atomic bombs were exploded, Babha said with remarkable prescience: 'When nuclear power has been successfully applied for power production, in, say, a couple of decades from now, India will not have to look abroad for its experts but will find them ready at hand.'

From 1945 onwards, some Indians saw atomic power as a means of advancing India economically, and also of enhancing its status as a big power. India has the world's largest deposits of thorium, as well as much smaller ones of uranium. Thorium is a naturally radioactive element that can be made fissionable by bombardment with neutrons. Like uranium 238, it is 'fertile', which means that it is not fissionable but can be made fissionable. It was thought at one time that thorium would play a larger part in nuclear energy than it has played; the Acheson-Lilienthal Plan in 1946 called for thorium as well as uranium deposits to be placed under international control.

India took up enthusiastically the IAEA's early optimistic forecasts about the benefits of nuclear energy in the poorer countries. Thus, Homi Babha, in 1954, said: 'For the full industrialization of under-developed countries and for the continuation of our civilization and its further development, atomic energy is not merely an aid: it is a necessity.' A few people said it was not a necessity as a power source for India since India has large coal deposits and a railroad system capable of transporting it, but they were shouting against the *zeitgeist*.

Babha was elected Chairman of the 1955 UN-sponsored Atoms-for-Peace Conference, and stated again his commitment to nuclear energy in his report to the General Assembly: 'Even if the widespread use of atomic energy for peaceful purposes raised political and military problems, there would be no option but to solve those problems.' (The syntax is telling, implying as it does that 'political and military problems' are only a hypothetical possibility.)

Nehru was persuaded of the importance of nuclear power for India very early on. He established an Atomic Energy Commission in 1948, with Babha at its head, and then, in 1954, a Department of Atomic Energy, which was responsible directly to the Prime Minister. Nehru and Babha were very close, and addressed one another in letters as *B'hai*, a Hindi term of intimate address that means 'brother'. In his acquaintanceship with the corridors of power, Babha reminds one somewhat of Robert Oppenheimer, and in other facets also: his brilliance as a scientist; his ability to gather other scientists around him and to lead them; and the wide range of his interests and tastes – Babha was an accomplished violinist, and the walls of his house in Bombay were hung with abstracts that he had painted himself.

Indian scientists built Asia's first nuclear reactor, a 1-megawatt research reactor, and it went critical in 1956. It operated on enriched uranium given by Britain.

During the Cold War of the 1950s, India took a position of

neutrality. This was condemned by John Foster Dulles, for whom the Cold War was a struggle between the forces of light and the forces of darkness. But many people found the stance preferable to the belligerent righteousness of the major contenders. There was at this time a kind of mystique about the newly independent non-white countries among liberals in the West. They were presumed both to be free of many of the sins of the big powers, and also to be owed a debt because of their past colonial subjection. India was the principal recipient of these feelings, as the largest such country and a democracy to boot, a country which at its birth bore the imprint of Mahatma Gandhi.

On atomic energy, Nehru said in August 1957: 'India will in no event use atomic energy for destructive purposes, but only for peaceful purposes. I am confident that this will be the policy of all future governments.' For many people outside India, this hardly needed saying. Nehru seemed to stand for a higher morality than nuclear deterrence. Indian leaders had often expressed their abhorrence of nuclear weapons, and Nehru sent messages of support when the Campaign for Nuclear Disarmament staged its big rallies. After Russia and America both resumed nuclear weapons testing in 1962, Nehru addressed an Anti-Nuclear Arms Convention in New Delhi attended by people from many countries, which proposed vigils against nuclear war in the capitals of all the nuclear weapons states. Another participant in this conference was Mrs Indira Gandhi.

Canada at this time had a very activist External Affairs Minister, Lester Pearson, and he sought a greater role for Canada in world affairs. As Canada's ambassador to the United Nations, he had initiated important moves on nuclear arms control. Now he felt that Canada could help give more substance to the British Commonwealth, uncommitted as an organization in the Cold War, and standing for multi-racial co-operation and peaceful development. Canada extended aid to India under the Colombo Plan for Commonwealth development. As a part of this aid programme, Canada signed an agreement in 1956 to provide India with a 40-megawatt research reactor, of CANDU design, called CIRUS. It went critical in 1960. For many Canadians, aid to India in advanced technology symbolized all that was positive and healthy in their country's role in the world. This helps to explain the very emotional reaction to the use of this aid to create a nuclear explosion.

During the 1960s, Canada helped India build two 220-MW CANDU power reactors, called RAPP 1 and 2. India also bought two slightly smaller PWRs from Westinghouse, and these were installed at Tarapur, near Bombay. But this was a deviation from the mainstream of the Indian nuclear power programme. This programme was based in the first place on CANDU reactors. There was good reason for this. India aimed at nuclear self-sufficiency. The CANDU reactor uses natural uranium as fuel, and India has deposits of uranium in the sand on Kerala's beaches. The PWR uses enriched uranium; India has no enrichment plant, so the fuel for these has to be bought abroad. All these reactors were placed under the safeguards system that was now the norm in international nuclear trade. But CIRUS was started before these safeguards came into operation.

India's plan for self-sufficiency consists of three phases. In the first phase, it has natural uranium reactors producing power. In the second phase, it is to have breeder reactors, which will use plutonium produced in these earlier reactors, but somewhat different from the standard breeder reactor. The ordinary breeder has a blanket of ordinary uranium around the fuel which is converted by the fission into plutonium: these breeders will have a blanket of thorium, and the thorium will be converted by the neutron bombardment into U-233. In the third phase, the breeder reactors will burn a mixture of thorium and U-233.

India's insistence that it had no intention of building nuclear weapons always went hand-in-hand with extreme sensitivity to any inspection of its nuclear facilities. India always looked on the IAEA's system of inspection with great suspicion, and talked about the dangers of 'economic neo-colonialism.' Babha feared that international controls might restrict India's development. 'We should not make the mistake of putting a new-born infant into chains in order to ensure that he will never grow up into a criminal,' he said. India never ceased to denounce the Non-Proliferation Treaty because of the way it discriminated between nuclear weapons states and non-weapons states. The ever-quotable Babha declared: 'The NPT would cage the puppy of horizontal proliferation while leaving the tiger of vertical proliferation free to roam the world.'

The 1956 agreement with Canada on the supply of a CANDU research reactor contained no provision for inspection. It simply said: 'The Government of India will ensure that the reactor and any products resulting from its use will be employed for peaceful purposes only.' After this, the practice of inspection of nuclear plants by a donor country was established, and India had to allow Canada and the United States to inspect the reactors they sold to India, as a condition of the sale.

In pursuit of its long-term plan for nuclear power, which called for breeder reactors fuelled by plutonium, India built a reprocessing plant, to reprocess used fuel and extract plutonium, and this was completed in 1964. It is housed in a rectangular concrete building about as tall as a four-storey house at the nuclear research centre at Trombay.

Today, there is international alarm when any new country makes moves to acquire a reprocessing plant, and the big powers exert diplomatic pressure to head it off. But no one expressed any concern when the Indian plant started operating, even though India would not need plutonium for a breeder reactor for many years (an Indian breeder is *still* a long way in the future).

People who knew Babha say he always wanted to keep open the option to build a bomb, and ensured that studies on nuclear explosives were continued. On the day after the 1956 agreement with Canada was signed, Babha had breakfast with a visiting Australian scientist and told him that he wanted to stage an explosion as a scientific experiment. Nehru would not hear of his doing any work towards a nuclear explosion. But there are signs that his opposition weakened after the border war with China of 1962. This was a great blow to him personally, since he had promoted the policy of friendship with China. After 1962, most Indians felt that there was a potential enemy across the Himalayas, and in 1964 it became a nuclear-armed one.

After Nehru's death in 1964, Babha drew up a plan for an underground nuclear explosion (India had signed the ban on nuclear tests in the atmosphere) and gave it to the new Premier, V.S. Shastri. Shastri showed no interest until the war with Pakistan in 1965 over Bangladesh. This left India dominant militarily, but the American support for Pakistan worried the Indian Government, particularly the warning gesture of sending the aircraft carrier *Enterprise* into the Indian Ocean. Indians have always talked of their nuclear explosion abroad as a peaceful one in intent, but many of the factors that appear to have influenced developments are strategic ones.

Shastri approved preliminary work on an underground explosion in December 1965, a decision that was never announced. But Shastri gave an interview to the *Guardian* in which he said that an atomic explosion was not necessarily an atomic weapon, and this was taken widely as a hint that India might be planning some such thing.

Lorne Grey was worried by this talk, and he told Babha he wanted to see him to discuss the question of a 'peaceful' explosion. Babha was planning to go to an IAEA meeting in Geneva, and then on to the IAEA headquarters in Vienna, so they agreed to meet in Vienna. Babha left Bombay for Geneva on an Air India Boeing 707 on 25 January 1966. The weather over Switzerland was unusually bad. In freezing temperatures and fog, the airliner crashed into the side of Mont Blanc, and all the passengers and crew were killed.

The day of the crash was also the day that Mrs Gandhi took office as Prime Minister. She cancelled the plan for an underground explosion. But the possibility remained, and it was talked about within the Indian nuclear establishment. Just as Indians had once taken to heart the IAEA's over-optimistic forecasts about the benefits of nuclear power, so they now took up the idea that had been aired in America and Russia, that there were scientific benefits to be gained from observing a nuclear explosion, or at least, they affected to take it up.

In January 1971, Pierre Trudeau, the Canadian Prime Minister, paid a visit to New Delhi, and discussed proliferation questions with Mrs Gandhi, among other things. By October of that year, he was sufficiently disturbed by the talk of nuclear explosions to write to Mrs Gandhi:

> You will remember that in our talks, I referred to the serious concern of the Canadian Government regarding any further proliferation of nuclear explosive devices. The position of my government on nuclear explosions has been stated on a number of occasions, and you will no doubt be aware of it. The use of Canadian-supplied material, equipment or facilities in India, that is, at CIRUS, RAPP I or RAPP II, or fissile material from these reactors, for the development of a nuclear

> explosive device would inevitably call on our part for reassessment of our nuclear cooperation with India, a position we would take with any other non-nuclear weapons state with which we have cooperation arrangements in the nuclear field.

Mrs Gandhi replied: 'The obligations undertaken by our two governments are mutual, and they cannot be unilaterally varied. In these circumstances, it should not be necessary now, in our view, to interpret these agreements in a particular way based on the development of a hypothetical contingency.' This was fairly opaque, but the Canadian Government took it as meaning that the question of a nuclear explosion need not arise. It was on the basis of this letter that it accused Mrs Gandhi later of bad faith.

After this, there were hints from Indian officials that India was contemplating a test explosion, and the US and Canadian governments made several statements to the effect that so far as they were concerned, there was no difference between a peaceful and a military explosion. Meanwhile, public opinion polls in India showed a clear majority in favour of India having nuclear bombs, some opposition members of Parliament called for a bomb programme, and the Government found itself under increasing pressure to do something to boost national morale.

It seems likely that work towards a test explosion started before a firm decision was taken. One scientist involved said later, according to an article in the Indian magazine *Science Today*: 'Not much was written down. Like the Vedas, we carried on mostly by word of mouth. Though many people given specific jobs were not told what they were doing, many sensed it all the same. But they never mentioned it. It was a kind of tacit agreement.'

The final decision was taken in Feburary 1974, and great care was taken to maintain secrecy. One group of scientists within the AEC did the work, and all knowledge of it was kept from the others. The site was in the Rajasthan Desert, near the Pakistan border, where there was usually a good deal of activity anyway. The two CANDU power reactors and the PWRS were under safeguards; Canadian and American inspectors checked to see that the fuel was not reprocessed, or if it was, that all the plutonium was retained. But there was no check on CIRUS, the 40-megawatt CANDU research reactor. The spent fuel rods from this were reprocessed, and plutonium extracted to make an explosive device. The CANDU uses heavy water as a moderator, and this was supplied by the United States. The Chairman of the AEC, H.N. Sethna, took charge of the operation personally, and arrangements were made to send a message from the test site to Mrs Gandhi as soon as a successful explosion took place. The signal was to be: 'The Buddha smiles'.*

In India, the news was almost universally welcomed. Members of Parliament, the Press and the public seemed equally elated. It was an immediate boost for Mrs Gandhi's Government. A national railroad strike was threatening to paralyse India; now Mrs Gandhi used her executive powers to order the strikers, as public employees, to go back to work, and they went. Government and public alike seem to have been unprepared for the strongly adverse reaction abroad. Before turning to this, one might consider some of the lessons that can be learned from this episode.

One is that any kind of a reactor plus a reprocessing plant is a facility for making nuclear explosive material if anyone wants to use it for that. This was already realized by the time of the Indian explosion, but it was not appreciated earlier, when the operation or construction of these facilities was encouraged as a sign of modernization.

Another is the durability of hardware and the relative impermanence of governments. The CANDU reactor and the reprocessing plant outlasted Nehru and his era of Indian politics. In 1978, when the United States was trying to impose new anti-proliferation conditions on the supply of nuclear fuel to European countries, one Belgian said to an American official angrily: 'Are you really worried about us? We're not Libyans! We're not Indians!' One answer to this might have been that the Indians were not Indians either when the CIRUS reactor was handed over, or at least, not today's Indians.

---

* The business of nuclear explosives seems to turn ears to tin. To anyone with even the slightest acquaintance with the teachings of Gautama Buddha, the coupling of his name with a nuclear explosion must seem grotesque. A similar insensitivity was seen in the signal that told Truman that the first atomic bomb was tested successfully: 'Baby born on schedule'; most people's reaction to the arrival of this thing in the world were not those that greet the birth of a baby. And the US AEC termed its first study of strontium 90 radioactive fallout from nuclear bomb tests Project Sunshine.

Another lesson is that it is not enough to ask whether a country with nuclear power facilities is planning a nuclear explosion. A country is not a monolithic entity. It may well be that a cabinet has no such thing in mind, but that a small number of people, such as a group of government scientists, are keeping the option open. A group of military men could perfect the non-nuclear mechanism for a nuclear bomb without the public or even the government knowing about it.

Another is that it is quite easy to take up this option once the fissile material has been acquired. As Roberta Wohlstetter wrote in *US Peaceful Aid and the Indian Bomb*: 'With cumulating changes that shrink the critical time, only a minor event is needed to tip the decision in the timing for exploding a nuclear device: for example, a mere 'tilt' towards Pakistan by the United States rather than a reversal of alliances, or the need for distraction from transient domestic economic troubles such as a railroad strike.'

The British and Canadian governments both issued statements deploring the Indian explosion. The US Government said little, and made no immediate changes in its trade policies towards India. Later, there would be much agonizing and arguing in Washington over nuclear exports to India, but in the years after the explosion, America continued to send fuel to the reactor at Tarapur, the only one that uses enriched uranium. After all, this reactor was under American inspection to ensure that nothing emerging from it was used to create an explosive. Nor did any Western countries cut their economic aid to India.

Some alarm was expressed among some of India's neighbours, but they were concerned with the balance of power, not with the principle of nuclear proliferation. The big powers were worried that one more nation might have joined the nuclear club; they were worried only that this country was India. China, which had already forged its links with Pakistan, offered to guarantee the defence of any country that was subjected to Indian nuclear threats (just as, ten years earlier, when China exploded its first nuclear bomb, President Lyndon B. Johnson offered to defend any country threatened by China). India continued to insist that it did not have an atomic bomb nor any intention to build one.

British and Canadian officials remonstrated privately with

Indian officials and told them that India could ill afford to spend money on nuclear explosions. The Indians replied that the explosion had not cost much: they estimated the cost of all the materials used in it, including the plutonium, as under 300,000 dollars. This, they pointed out, was rather less than the Babha Atomic Research Centre earned in a year from the sale of radioactive isotopes to industry and to hospitals in India and abroad. This was a specious argument since the cost of the nuclear explosion was not merely the cost of the nuclear materials; a lot of the Indian nuclear energy programme made no sense unless it was related to the explosion, or at any rate, it has produced no other tangible results. Press commentators carried the argument further, and said that Britain, as a nuclear weapons country, was in no position to counsel others to abjure even peaceful nuclear explosions, while Canada could afford to take a lofty view because it sheltered under the American nuclear umbrella.

The Canadian Government cut off all aid to the nuclear power programme. It had supported the CANDUs with advice and some of the sophisticated components that might be a part of any advanced industrial technology, such as refined magnets and pressure valves. It also cut off the supply of heavy water, which is used as a coolant in the CANDU.

Two senior officials flew out from Ottawa to try to find terms on which nuclear aid or at least nuclear trade could be resumed, Ivan Rowe from the Prime Minister's office and Michel Dupuy from the Department of External Affairs. They wanted the Indian Government to accept international inspection of the entire nuclear programme – full-scope safeguards. The Indians refused. At one point, Rowe and Dupuy thought they had reached an acceptable compromise. But when they got back to Ottawa, they found their terms were unacceptable to the cabinet, and they had to tell the Indians this.

Sethna was even angrier now that he had been at the original cut-off, and all the more so when the US Administration began threatening to stop the supply of fuel to Tarapur, which helps light the lamps of Bombay.

He got heavy water from Russia, albeit under safeguards so strict that it could never contribute to a nuclear weapons programme, and

Indian engineers learned to make the other components that Canada had supplied. Five years later, in 1979, Sethna was to write in an article in the IAEA Bulletin:

> India's programme for the peaceful utilization of atomic energy has, to a certain extent, been affected by restrictive trade practises and unilateral trade embargoes on nuclear supplies by certain countries. . . . Such measures will cause temporary delays and perhaps cost over-runs for projects in the near future. In the long run, these developments will only strengthen and accelerate India's self-sufficiency in the nuclear field.

Mrs Gandhi's dictatorship ended in 1976 and Morarji Desai was elected Prime Minister. Pressed now by the US Government as well, he expressed public regret for the explosion, and said it had harmed India. But he also would not accept full-scope safeguards. He made a simple offer: he would accept them if the superpowers did also. Since this would apply to all nuclear facilities, military as well as civil, it would mean that the superpowers could not manufacture any more nuclear weapons, and this they clearly would not accept. As for signing the NPT, he said again that it was easy for Canada to do so in her sheltered position, but that no non-aligned country which had the capability to make nuclear weapons had signed it.

Later still, in 1979, Mrs Gandhi was elected Prime Minister again, and she took a stronger line. US-enriched uranium was still being sent to Tarapur, but each shipment was the subject of debate in Washington and required special permission of Congress. Because of this uncertainty about fuel supplies, the two reactors at Tarapur were operated at a power level far below their capacity. The Indian Government evidently found the position constraining. In early 1981, it decided that it would no longer remain on the good behaviour required of it to try to qualify for American approval. Sethna said it would reprocess used fuel, extract the plutonium, and use it mixed with uranium as fuel for power reactors. Steps were taken to terminate the 1963 agreement with the United States on fuel supply.

Meanwhile, Pakistan took steps in the direction of building nuclear weapons, and Mrs Gandhi warned in Parliament in April 1981 that if Pakistan exploded a nuclear device, India would respond 'in an appropriate way'.

Chapter 7

PLOWSHARES AND SWORDS

Since the Indian explosion, nuclear proliferation has been seen in a new light. It is not simply the spread of nuclear bombs, but the spread of potentially explosive fissile material, or the means to manufacture it, with or without safeguards. The change of attitude was not really as abrupt as this: a number of people had worried publicly for some time about the dissemination of nuclear technology, for instance, Leonard Beaton, in his widely noticed book *Must the Bomb Spread?* But the Indian explosion dramatized the dangerous potential, and focused the attention of Government officials and many in the media in the direction in which Beaton and others had been pointing. From now on, the issue of proliferation was linked to the presence of enrichment plants and used fuel reprocessing plants, from which could issue forth highly enriched uranium and plutonium. Because of this, it is linked also to a number of other issues: the scarcity or abundance of uranium; the reliability of the United States as a fuel supplier; the disposal of radioactive waste; and the efficiency of a breeder reactor.

There has been a concentration on plutonium in particular, partly because India took the plutonium route to a nuclear explosive; partly because plutonium is going to be plentiful as more and more reactors are built; and partly, perhaps, to compensate for the fact that insufficient attention had been paid to it before. Techniques for enriching uranium remain a tight secret, but the United States told the whole world how to reprocess used fuel and extract the plutonium, in several thousand technical papers declassified at the two atoms-for-peace conferences.

This new anxiety about the spread of some nuclear technologies coincided with anxiety about how nuclear power industries could survive economically. One answer to this was to export, and sometimes to export the same technology that was causing the concern about proliferation. Commercial requirements came into conflict with the requirements of an anti-proliferation strategy, and they do still.

Of course, no one sets industrial profits in the balance against the incineration of cities openly, nor even, one imagines, consciously. The issue is never posed in such stark terms. It is posed in terms of what moves on exports may have what long-term effects, whether helping some countries acquire certain weapons-sensitive facilities might encourage others to acquire them also, and the consequences of a more permissive or more restrictive attitude to nuclear technology.

For a while, the prevailing attitudes to nuclear power remained strongly positive. Flagging enthusiasm was given a boost by the dramatic oil price rises of 1973–4. After this, no one built any more oil-burning power plants.

The French Government reacted to the oil crisis by making the biggest commitment that any government has yet made to nuclear power. This became central to the long-term strategy of industrial expansion and rapid advance in new technologies. The French Government believed that this could not be achieved while France was almost totally dependent for her energy on imported raw materials. The key role of nuclear power in French plans was seen in the appointment of the Director of the CEA, André Giraud, as Minister for Industry. 'Our country,' Giraud said shortly after his appointment to the cabinet, 'has no serious alternative to nuclear power except economic recession.

The nuclear power programme was ambitious: nuclear energy was to supply 50 per cent of France's electricity by 1985. The Government ordered thirteen 900-MW nuclear power plants, the largest order of this kind ever placed anywhere. These were all to be manufactured by one company, Framatome and all to be delivered to Electricité de France, the nationally-owned power company. The French Government pushed ahead with the programme with characteristic energy and singlemindedness. Public debate was not encouraged. When a radio interviewer put it to Giraud that there might be increased public involvement in decisions, he said, 'How do you expect the public to find its way when some scientists contradict other scientists? The only thing to do is to carry on with our work, and then persuade people that nuclear power affairs are in the hands of serious people who deserve their confidence.' The first reaction to Press reports that there were leaks in some reactors was to clamp down on leaks to the Press. The programme continued at full throttle until the election of Francois Mitterand, who took the presidential foot off the accelerator.

The Giscard Government made one important policy change. It decided to abandon its gas-cooled reactor design and switch to PWRS, which Framatome would build under licence from Westinghouse. The better performance of the PWRS and attractive purchase terms persuaded the authorities, even though this meant relying on America for the supply of enriched uranium fuel.

The Soviet authorities, ideologically committed to technological optimism, also pressed ahead with their nuclear power programme. Scientists and officials seemed confident that nuclear power could supply a large part of their country's electricity, and that breeder reactors would solve the nuclear fuel problem forever. They started with graphite-moderated reactors but then switched to PWRS similar to the American one in design. A huge nuclear power complex sprang up at Novovorenzh, 300 miles south of Moscow on the Don River, with five nuclear power plants, other nuclear facilities and industrial plants drawing power from those, and a town of 25,000 people.

The Russians were so confident on safety that they built their reactors without the thick concrete containment structures that enclose nuclear power plants elsewhere and serve as a backstop in case of a radiation leak, because, they say, an accident is impossible. Now they are putting concrete containments structures around their newest ones, though some Soviet scientists have told Westerners these are quite unnecessary and a waste of money. When a serious accident did occur in 1959 in the Southern Urals with a release of radioactivity, admittedly at a military establishment, the authorities simply suppressed all information about it, so that it has come out only recently, and the authorities will still not admit to it.*

Germany also, with dwindling coal deposits, decided to rely heavily on nuclear power for its industrial future, and it accepted

Zhores Medvedev, the Soviet biologist now living in Britain, believes there was a burst of radioactivity from nuclear waste which killed and injured many people and contaminated a considerable area of land, and has written a book recounting it in these terms, *Nuclear Disaster In The Urals.* Others examining the evidence are less sure of the nature or extent of the accident, but there is strong evidence that something happened and that it caused loss of life and contamination.

the light water reactor as the standard power plant. Both types of LWR were built, but eventually Kraftwerk Union (KWU) emerged as the reactor-building consortium, and it dropped the BWR and concentrated on the PWR. Few Germans had any qualms about relying on the United States for nuclear fuel. Two generations had grown up accustomed to relying on the United States for their national defence.

The US Atomic Energy Commission itself created the first crack in other countries' confidence in the United States as a supplier of fuel. First, in October 1974, it became worried that it might be over-committing its enrichment facilities, and it imposed stringent conditions on new contracts, designed to discourage would-be purchasers of enrichment services. Orders must cover a minimum of eight years, and part of the fee must be paid in advance. Then, in March 1975, it announced that it was suspending all licences to export uranium. It said it would review the situation, and issue licences only if it was sure there was enough enrichment capacity free to meet domestic American needs.

The AEC did not consult its customers before this announcement was made, and it came as a blow. The European Commission protested formally on behalf of Euratom. It pointed out that the provision of enriched uranium was a part of the US Government's agreement with Euratom, and declared itself 'astonished' that the United States would break this agreement unilaterally.

In retrospect, it does indeed seem astonishing that the AEC should have made this announcement, since it turned out not to be necessary to halt any shipments of enriched uranium. The unexpected announcement was almost as great a shock as an interruption would have been.

It gave a slight extra push to the feeling that it might not be wise to rely on the United States exclusively for enriched uranium, and that Fritz Efelfurt may have been too sanguine when he assured other members of the European Commission that they would be able to buy enriched uranium just like coal. Now several European countries were building their own civilian enrichment plants. Britain, which was already enriching fuel for its atomic warheads, built an enrichment plant at Capenhurst, Cheshire, to enrich uranium for its AGRs. France joined forces with Belgium, Spain and Italy to form Eurodif, which built a huge gaseous diffusion plant at

Tricastin, on the banks of the Rhône near the military enrichment plant at Pierrelatte. Iran later joined the project, and later still dropped out, when the post-Shah government abandoned nuclear power. This plant only began operating in 1979, and even then, it was not able to fill all France's needs.

West Germany, meanwhile, found an unexpected opportunity to diversify its sources of reactor fuel. The German Government abandoned its totally negative policy towards Eastern Europe in favour of the Ostpolitik. Discussions took place with several Communist bloc countries about sharing electric power production or transmission. The West Germans saw political dangers in this and dropped the idea. But in the course of discussions with the Soviet electrical power authorities, it emerged that the Soviet Union had spare uranium enrichment capacity (an indication that its reactor-building programme was lagging behind schedule), and would be willing to take orders for enrichment. Despite some official misgivings, Germany signed a contract with Russia for uranium enrichment in 1975.

Soon, Russia was supplying 45 per cent of Germany's enriched uranium reactor fuel, and it has now begun selling to other Western European countries as well. This has meant a shift in the trade routes. Germany now buys Canadian uranium and, instead of sending it all south of the border to be enriched in American plants before shipment across the Atlantic, it sends some of it to Russia. The Soviet authorities have not told any Westerners details of their enrichment plant, not even where it is located. They simply collect the raw uranium, and deliver it enriched to West Germany.

Meanwhile, another method of enriching uranium, the centrifuge method, was being developed in Germany. It originated in the work of a German chemist called Hans Zippe, and he developed it originally in the Soviet Union. Zippe found himself in the Soviet occupation zone when the war ended, and, like many German scientists, he was taken to Russia to work on a scientific project, in his case, the nuclear power programme. He drew up plans for a machine to enrich uranium by centrifuge, and later, when he went to West Germany, he took out a patent on it.

As with gaseous diffusion, the principle of centrifuge is easy to understand, the engineering very difficult to accomplish. The uranium is converted to a gas, uranium hexafluoride, as it is in a diffusion plant. The gas is whirled around and around so that the heavier U-238 atoms are flung outwards, and the gas near the centre contains an increased proportion of U-235 atoms. The engineering skill lies mainly in building rotors powerful enough to attain the required speed and components strong enough to withstand the strain. The rotors perform 50,000 revolutions a minute, and when they are doing so, the outer wall of the centrifuge is moving at the speed of sound. Zippe's first patent was for the ball bearing base, which is a vital component. A centrifuge plant consists of hundreds of such identical machines; the gas that collects near the centre is passed from one machine to another, so that the proportion of heavier atoms is increased each time.

A centrifuge plant costs more to build than a gaseous diffusion plant, but costs less to operate because it requires less electric power. This is an important saving; uranium enrichment is still the most expensive process in the nuclear fuel cycle. A centrifuge plant can be operated economically at only half capacity or less if there is a small amount of uranium to enrich, whereas a gaseous diffusion plant can only operate at full capacity whatever the amount of uranium, and the capacity of the smallest gaseous diffusion plant is enough to supply fuel to five 600-MW nuclear power plants. The cost advantage is the reason that the huge new enrichment plant that the US Government is building at Portsmouth, Ohio, is to be a centrifuge plant.

There is another difference which has far-reaching implications. In a civilian gaseous diffusion plant, uranium cannot be enriched to a high enough degree to use in an atomic bomb. The chambers between the porous membranes are large, and several pounds of uranium hexafluoride collects in each of them. If uranium in this gas were enriched so that it was more than 50 per cent U-235, then it might start to fission spontaneously, and one would have a chain reaction with a large release of radioactivity. One could dismantle and reassemble a civilian diffusion plant so as to enrich uranium to weapon-grade, but this would mean constructing much smaller chambers and making other alterations, which would be difficult and very visible.

A centrifuge plant, however, can enrich uranium to any degree: only a small amount of uranium hexafluoride is collected in each machine in any case. It is not possible to tell from the outside whether it is enriching uranium to $3\frac{1}{2}$ per cent for reactors or to 98 per cent for bombs. (It is possible to tell from inside the plant, however, at least for someone familiar with the process. Sizeable amounts of uranium hexafluoride collect in the tubes which carry it away from the many centrifuge machines to be liquified, and the tubing would have to be modified to keep this amount down.)

Because of this particular characteristic, the Western allies let the German Government know that they would not like to see a wholly German-owned centrifuge plant. So Degussa AG, the big German chemical and manufacturing company that bought Zippe's patents, joined with others. In 1971, the West German, British and Dutch governments signed the Treaty of Almelo, agreeing to pool their resources to develop centrifuge enrichment. The result is Urenco, a tri-national organization which has centrifuge plants in all three countries.

A new trade in reactors was growing up. France and Germany, having acquired the technique of making light water reactors from American companies, were now competing with their teachers as exporters. Also, Canada was still trying to sell its CANDU abroad, and Britain to follow up its two sales of Magnox reactors to Italy and Japan with others.

The competition to export was fierce because the nuclear power industry was losing money. The mid-1970s were a time of setbacks for nuclear power, or perhaps one should say the beginning of a time of setbacks, for they have not stopped. A lot of shining hopes of a few years earlier were shrivelling.

Production costs were much higher than the manufacturers had foreseen, and they rather than the customers were paying them. The environmental movement was beginning to bite, with new regulatory requirements, and safety hurdles were increasing both manufacturing and operating costs. General Electric, which had made such confident forecasts about Oyster Creek, lost $600 million on its first thirteen reactors. Westinghouse and Babcock and Wilson lost big sums on reactor manufacture. General Atomics, one of the first American companies in the field, owned by Gulf Oil and Royal Dutch Shell, went out of the nuclear power business. In Germany, AEG-Telefunken, one of the two partners in the reactor company KWU, sold its share to the other partner, Siemens, at a loss and pulled out.

Orders fell off. In the United States, in 1975, five new nuclear power plants were ordered and eight existing orders were cancelled.

The industry needed export orders to survive. A top executive of KWU, Gunter Hildenbrand, wrote in an article that the development and manufacture of nuclear power plants is only viable if there is an export market equal in size to the domestic market.* In Britain, the Chairman of the Atomic Energy Authority, Sir John Hill, warned that the British industry was in danger of collapse unless 'a vigorous export policy was pursued'. (It was not, or at least, not successfully; no one wanted to buy gas-cooled reactors any more.) It was a buyers' market, in which buyers were in a position to set the conditions of sale.

* * *

Despite the disappointments in the performance of nuclear power, its benefits to the Third World were still being over-stated in the industrialized countries. The IAEA did a survey in 1974 to see which countries could benefit from nuclear power if the viable size of nuclear power plants were brought down to 150 MW, and another if it were brought down to 100 MW. There were forty-six countries on the first list, including Cuba, and sixty-one on the second, including Uganda. The then Deputy Director-General of the IAEA, David Fischer, says now: 'These lists look pretty silly today. And pretty frightening too, I suppose.'

Naturally, the Third World countries that were most interested in acquiring nuclear power were those with a fast-growing industrial sector. Some of these were prepared not to look too closely at the economic competitiveness of nuclear power, because they were keen to be involved in the newest high technology. Some of them wanted to buy not just nuclear reactors, but other facilities with which they could make their own fuel, and ultimately be independent of outside sources of supply. In the wake of the Indian nuclear explosion, the United States, the self-appointed guardian of the non-proliferation régime, became alarmed at the prospect of these sensitive technologies spreading, and moved to prevent it, with the support of some other countries.

---

* *International Security*, Fall, 1978.

South Korea, which was already building three nuclear power plants, decided that it wanted a reprocessing plant, to reprocess the used fuel and extract the plutonium for future use. France's CEA offered to build one. The United States, mindful of the use that India made of its reprocessing plant, scotched that plan. It expressed its disapproval to the Seoul Government, and when there seemed a chance that this would be insufficient, it followed it with a strong, secret warning that if South Korea went ahead and acquired a reprocessing plant, then the United States would have to reconsider their defence relationship. This did the trick.

The Shah of Iran, with his grandiose programme of industrial expansion and modernization, wanted Iran to have a network of nuclear power plants and also enrichment and reprocessing plants as well. The Shah negotiated with France and Germany to buy these. The United States, then Iran's principal supplier of military technology, persuaded the Shah to make the projected reprocessing plant multi-national, and for enrichment, to settle for a share in Eurodif instead of building a national enrichment plant.

The United States did not have the same leverage everywhere. Two deals in particular worried the US Administration, and worried others as well. The CEA had signed a contract to build a reprocessing plant for Pakistan; and KWU in Germany had landed a massive order to set up in Brazil a whole nuclear power network: eight nuclear power stations, plus pilot enrichment and reprocessing plants.

All these plants were to be under IAEA safeguards, but this did not remove anxieties for all time. Pakistan in particular was worrying. There seemed no sound economic reason for building a reprocessing plant there now. Pakistan had only one power reactor, a 500-MW CANDU supplied by Canada. But the reprocessing plant would have the capacity to handle fuel from eleven nuclear power plants. When this was pointed out to Pakistani officials, they said only that Pakistan had a national plan to build twenty-four nuclear power plants by the year 2,000 and breeder reactors also, and they wanted Pakistan to have a reprocessing plant that could supply plutonium for the breeder comfortably in advance.

The United States approached both Germany and France to call off these sales. This was a very different world from the one of a few years earlier, when Germany was still out to show itself a loyal ally and good European with the old nationalist devil buried, while

Gaullist France was stubbornly nationalistic and chronically anti-American. Now France proved the more pliable of the two, while Germany stood out vigorously and unashamedly for its national interest; this stance was more acceptable internationally with a Social Democratic like Helmut Schmidt as Chancellor than it would have been with a Christian Democrat.

At first, the French were annoyed at American objections to the Pakistani deal. They themselves had felt some doubts about it, partly because they feared that it would upset the CEA's good relationship with India. When they agreed to sell a reprocessing plant, they insisted on IAEA safeguards, something France had never done before (Pakistan had not signed the NPT). President Ford himself became sufficiently involved to press his concerns on President Giscard d'Estaing when they met at the four-power summit meeting on the French Caribbean island of Guadaloupe in 1975, and Giscard was impressed. This conference was called to discuss international monetary matters, but the alerting of Giscard to the dangers of plutonium spread was one long-lasting result.

The CEA was accustomed to looking at the political aspects of France's own nuclear power programme, but it regarded nuclear exports as an aspect of trade policy rather than foreign policy, and talked with the Ministry of Trade more than with the Foreign Ministry. It considered that, like the Ministry of Defence with its promiscuous arms sales, it was doing its bit for the French economy with its exports. Giscard changed this after the Guadaloupe meeting. He set up an inter-departmental committee on nuclear exports with himself at the head. And when the physicist Bertrand Goldschmidt retired as the CEA's Director of International Relations, Giscard replaced him, not with another CEA man, but with a high flier from the Foreign Ministry fresh from the Washington and London embassies, Francois Bujon de l'Estang.

The French Government was loath to cancel the contract with Pakistan to build a reprocessing plant, because this might mar its trading reputation. French officials no longer defended the sale to Americans, but told them: 'We won't renege on a contract. But if you can persuade the Pakistanis to pull out, we won't complain.' Even this position was abandoned eventually. The economic justification for a reprocessing plant in Pakistan looked even less plausible after East Pakistan broke away to become Bangladesh, and Pakistan dropped its more ambitious nuclear power plans. The French Government asked Pakistan to accept joint French and Pakistan control of the reprocessing plant, and when Pakistan rejected this, it refused to go ahead with the contract.

Germany was a different matter. The German Government defended KWU's huge export order to Brazil as a matter of vital national interest. The youthful British Foreign Secretary, David Owen, appealed to Chancellor Helmut Schmidt to desist from these sales as a fellow Social Democrat, but went away disappointed.

Schmidt was questioned about it on American television by Martin Agronsky in April 1977, and said:

> Roughly speaking, it's true to say that we're exporting nuclear reactors to any country that wants to buy one, or even more than one. In Germany, it's a future industry, and it at present employs 100,000 people, skilled labor. We don't have the same chance, for instance, in the field of civil aviation, where the world is dominated by American corporations, nor do we have the same chance in the field of computers, where, more or less, the world market is in the hands of American corporations.

These last sentences suggest a theme that emerged often in these Transatlantic exchanges: resentment that America, dominant in so many areas of international trade, should try to hobble European industries in one of the few areas in which they could compete successfully. Some Europeans have always suspected that this, and not concern about the spread of nuclear weapons, was the real motive behind American urgings.

This suspicion was reinforced when the Germans found that while American officials were asking Germany to refuse to build an enrichment plant for Brazil for the sake of world peace, the big American corporation Bechtel was discussing building an enrichment plant with Brazilian officials. Some Germans refused to believe that the State Department did not know about these talks, nor could they accept American assurances that Bechtel was wasting its time because it would never have been given a licence to export enrichment technology. The State Department could not be so ignorant of what business was contemplating in this field, the Germans insisted, nor could a big corporation be so shortsighted.

American officials were in the uncomfortable position of having to protest that it was and one could.

At about this time, there came to light another instance of German nuclear exports in circumstances of questionable responsibility. It turned out that Germany had already sold uranium enrichment technology to South Africa, notoriously a non-NPT signatory, in 1974. This was kept secret at the time, and when it emerged and the facts had to be told, both parties fudged the details.

The technology involved is a new enrichment process, though it works by the centrifuge principle, usually called the jet-nozzle technique. In this, the uranium is liquified, not gassified, and projected through a nozzle in a curved path in such a way that the heavier and lighter atoms are separated by centrifugal force.

South Africa has a large research reactor, Safari, fuelled by 93 per cent enriched uranium, which it imports from the United States. It has bought two PWRS from the French company Framatone, and it has domestic supplies of uranium, but so far, no enrichment plant to enrich the uranium for use in its PWRS. Ucor, the Government nuclear power company, bought the rights to the jet nozzle process from the German company Steag.

The German Government seems to have been particularly anxious to push this sale through. Like any sale involving nuclear materials or techniques, it came up for cabinet approval. Several ministers questioned the wisdom of the proposed sale, on the ground that it could set back Germany's relations with black Africa. After some discussion, the matter was dropped. Then the Minister for Economics licensed the sale without telling the rest of the cabinet. The fact that he could do so without any repercussions indicates that he had the Chancellor's backing. The German Foreign Ministry did not confirm until 1977 that components for a jet nozzle centrifuge were sold to South Africa, and then it said that these components were not suitable for enriching uranium to weapons grade. However, foreign technicians who have seen the equipment in Germany deny this; they say that, whereas they are not certain that it will work on a commercial scale, there is no reason why, if it does work, it could not be used to make weapons-grade uranium.

The South African Government does not admit to this day that it is using the jet-nozzle technique in the enrichment plant that it is building. South African officials say theirs is a 'stationary wall centrifuge', in which the walls of the centrifuge are stationary while the uranium liquid circulates, and that this is a South African invention. The Prime Minister, John Vorster, underlined this when he spoke at the start of construction of the enrichment plant in 1975: 'That this achievement could be reached without any assistance from foreign countries', he said, 'inspires enormous confidence in the future scientific and technical development of our country.' But technical journals say that the stationary wall centrifuge is simply the jet-nozzle technique plus a few small innovations.

The plant began producing 45 per cent enriched uranium in April 1981 to fuel the Safari research reactor. The United States had stopped supplying highly enriched uranium for Safari because South Africa would not sign the NPT. The enrichment plant is not expected to be operating at full capacity before 1984.

* * *

The enrichment plant to be built in Brazil was to operate by the jet-nozzle technique. The KWU contract with the Brazilian Government was signed in Dusseldorf in June 1975 in the presence of Brazilian dignatories brought over for the occasion, with the lavish ceremonial appropriate to a major international treaty.

It was certainly a coup for the hard-pressed KWU. Many today would agree with the judgement of Jose Goldenberg, a leading Brazilian physicist and past president of the Brazilian Society of Physics, that the deal was 'the biggest over-sell of the century'. A number of prominent Brazilians expressed similar sentiments at the time, though usually less caustically. Three weeks after the contract was signed, a joint meeting of the Brazilian Society of Physics and the Brazilian Society for the Advancement of Science approved by acclamation a document that said: 'We express reservations about the fact that in a country where 100,000 megawatts of hydro-electric potential exists, it is deemed necessary to turn to nuclear power on this order of magnitude.'

Brazil has been ruled by a mostly military dictatorship since 1964. Its economy has grown rapidly. Its rulers want to turn Brazil into a world power, and they want to make it independent in every way possible. Dependence on oil from abroad was an obstacle.

Nuclear power could reduce this dependence.

Brazil bought a PWR from Westinghouse, and even this was criticized by some nationalists as 'truckling to colonialism', because of the dependence on America for fuel. The AEC's suspension of export licences for enriched uranium justified Brazillians' worst fears about this. Officials tell visitors that it was this that made them determined to go for nuclear independence. Brazil has since placed a large order for enriched uranium from Urenco, the European enrichment company. If a nuclear weapons option is in anyone's mind, no hint of this has been heard in public.

The decision to embark on a big nuclear power programme was taken by a small number of people in the Government. Few leading scientists were consulted, nor, it seems, were many leading figures in the nationally-owned electric power company, which will be using the product. Most of Brazil's electricity today is hydro-electric power, and there is more of this available, though most of it is far from the industrial centres. Brazil also has large untapped coal reserves, though there is disagreement about the quality of the coal. In its document justifying the nuclear power programme, the Government said hydro-electric potential is limited, but the electricity industry does not agree with its figures. Projections of industrial expansion indicate that Brazil may face a power shortage a few years from now. Nonetheless, in the enormous scope of the nuclear power programme on which Brazil has embarked, and the disregard of prudent advice, one can see a reflection in the technological sphere of an ambition for national grandeur.

The arrangement with Germany was designed to give Brazil not only a full nuclear fuel cycle, subject only to IAEA safeguards, but also all the technological knowledge and training needed for nuclear independence. It was essential to the scheme, from Brazil's point of view, that most of the plants were built in Brazil by Brazilians. One clause in the contract said that all components were to be manufactured in Brazil unless KWU could show that it could make them 15 per cent cheaper. Another specified the training that KWU was to train Brazilian physicists and engineers in nuclear technology. Brazil's leaders looked forward to the day when Brazil would not only not have to import nuclear technology, but would sell it to Third World countries, and use it as a bargaining counter, perhaps in exchange for oil. It has already discussed some such arrangement with Iraq, which supplies a third of Brazil's oil.

The first stage of the programme has hardly fulfilled their hopes. The estimated costs of the programme have tripled from $10 billion to $30 billion. The cost of nuclear electricity will be far higher than that of other kinds. The schedule of work has been slowed down. The entire programme was to be completed by 1990; it now seems likely that only two nuclear power plants will be working then, not eight, and that the enrichment and reprocessing plants will not be working either. It also seems likely that part of the programme will never be completed. With the increased liberalization the programme has come under informed criticism from scientists and Congress and in the Press. Critics are having a field day with the mistakes made already. The first nuclear power plant was started on unstable rock; half-way through construction, new supports had to be sunk at great expense. The container walls are not up to the German standard in thickness or strength. The main site for the nuclear complex is at Angra, midway between Rio de Janeiro and Sao Paulo, an area which turns out to be the location of two earthquakes in 1962 and 1967. Some of the students sent to Germany for training did not have the necessary academic background to benefit. It seems unlikely that Brazillians will be able to build much of the high technology equipment.

This is a country that has tried to live beyond its means technologically. Yet curiously, in another direction, Brazil's drive for economic independence has not only been successful, but has been innovative in a way that the anti-nuclear advocates of 'soft energy paths' would thoroughly approve. Brazil has gone much further than any other country in coverting sugar cane to alcohol and using this to power cars, either in gasohol or as pure alcohol in cars converted to its use. Brazil has an advantage in having a long-established and highly organized sugar cane crop, and the cane farmers had an incentive to go in for alcohol fermentation when the world price of sugar fell. Brazil has already shown that alcohol produced from sugar cane is less expensive and more nearly competitive with oil than most people thought it would be, and it has pointed the way to further improvements in conversion and use.

Chapter 8

THE CLUB AND THE GAMBLER

The fluctuations in nuclear power programmes over the years have been reflected in the uranium market. This market, tied as it is to future expectation rather than solid fact, has been very volatile; the price of uranium over the years has bounced up and down like the state of mind of a manic depressive.

During the 1950s and most of the '60s, the major powers were worried that they might be short of uranium, and they bought all they could get their hands on. Suddenly, in the late 1960s, everyone had enough. The nuclear weapons countries had sufficient stockpiles for their weapons programmes; nuclear power was not yet a big industry, and its appetite for uranium was small. There was an over-supply at the mines. The AEC, which had earlier imposed a ban on the export of American uranium so that none could leave the country, now moved in the opposite direction, and imposed what was in effect an import ban, to protect American uranium producers. It said it would no longer enrich for domestic use any uranium imported from abroad, and since all American reactors use enriched uranium, this ensured that no American reactor operators would buy foreign uranium. This was a heavy blow to overseas producers, since America was the biggest potential market, and it exacerbated their over-supply situation and pushed down the price further.

The price of uranium on the world market fell from $20 a pound to less than $6 a pound. This happened during a period when the cost of mining was going up, partly because of the introduction of new health regulations covering exposure to radon gas. Many producers found they were selling uranium at a loss. The biggest mines in Australia closed down. Some mines closed elsewhere; Elliot Lake in Ontario became a ghost town, its population fallen from five thousand to a few hundred. Canada's production of uranium, which was 12,000 tons in 1960, fell to 3,000 tons in 1968. It looked for a time as if the world's uranium industry might shut up shop. This was something the governments concerned could not allow to happen. Already the Canadian Government was buying uranium from Canadian producers and stockpiling it in order to keep them solvent.

Production was passing into the hands of the big corporations. In the late 1960s, there were still hundreds of separate mines dotted throughout the carnotite areas of the western United States, many of them worked by the men who had found them earlier, during the uranium rush days, with a few hired hands. In some, the heavy concentrations of uranium were mined out by now, and the work was far less rewarding. As the price fell, and the mining became more difficult, many of these miners could not withstand the losses they faced, and either abandoned the mines or sold them to big corporations. Many of the biggest mining and oil companies now moved into uranium, taking the long view that even if the price was low now, nuclear power was likely to be a major energy source of the future. Exxon, Gulf Oil, Getty Oil, Phillips Petroleum and Anaconda Copper became uranium producers during this period, and are today among the world's biggest. Getty also went into uranium in Australia, and Gulf in Canada. The age of the lone prospector had passed.

In Canada, Rio Tinto Zinc, the British mining company, bought out the Joubin-Hirshhorn interests at Elliot Lake and set up Rio Algom, a Canadian subsidiary. This began RTZ's involvement in uranium mining in many parts of the world, and also its expansion from a medium-sized British company to being the biggest multinational mining company in the world. The Denison Corporation, a big Canadian mining company, bought out all the other mines at Elliot Lake.

Governments were involved in the uranium industry everywhere. The RTZ move into Canadian uranium came about at the prompting of the British Ministry of Defence. A senior Ministry official suggested to Sir Val Duncan, the Chairman of RTZ, that since nuclear weapons were an essential part of the British defence programme, it would be a good thing if a British company owned some uranium. Sir Val flew to Canada and negotiated a purchase price for the mines of several million dollars, under the impression that the Ministry of Defence would get him government financing. But the man from the Ministry responded to his glad tidings by saying that no, good heavens, he had not meant that the Government had any money to put into uranium. For three days, RTZ faced financial ruin, because it was committed to the purchase and did not have the money. Sir Val flew back to Canada and was able to borrow the money from the Toronto Dominion Bank.*

On another occasion, the Canadian Prime Minister, Lester Pearson, vetoed a plan by Stephen Roman, the Chairman of the Denison Mining Company, to sell all Denison's uranium production for twenty-five years to France on political grounds. Since he knew Roman personally, Pearson called him into his office to tell him of his decision. Roman, a very hard-headed businessman who began life in Canada as a penniless Slovak immigrant and made a million dollars by the time he was thirty, did not appreciate the broad national considerations that lay behind the decision to torpedo his carefully worked-out deal. As he recalled the conversation with Pearson later: 'I called him a son-of-a-bitch and walked out of his office.' †

* * *

The entanglement of governments with the uranium trade, and of commercial considerations with political ones, produced extraordinary consequences in the affair of the international uranium cartel, the US Government's anti-cartel policies, and Westinghouse.

When the price of uranium plunged so low that producers were being driven out of business, an international agreement to divide up the dwindling markets rather than competing was an obvious move, and it must have occurred to several people. The suggestion was first mooted by Canadian Government officials in discussions with Canadian uranium producers. Some people in the industry suspect that the idea originated elsewhere, and was injected subtly into these discussions by Sir Val Duncan of RTZ. He had flown to Toronto and had a long talk with Stephen Roman of Denison, and they might well have hatched the plan between them; Sir Val had earlier been instrumental in getting a marketing and price agreement among zinc producers.

---

* This story is not found in print anywhere, but it was told by Sir Val Duncan's successor, Sir Mark Turner, albeit not to me. – NM.

† This story was told by Roman himself, and it *was* told to me – NM.

At all events, the Canadian Government set things in motion. It told the heads of the four Canadian uranium companies, including the Government-owned Eldorado Company, that they must arrange among themselves viable conditions for uranium production. Then it suggested to other uranium-producing countries that they meet to discuss the critical situation. The meeting was held in Paris in February 1972, in the offices of the Commissariat à l'Energie Atomique, a six-storey office building near the foot of the Eiffel Tower. Actually, two separate meetings were held simultaneously: one of twenty-three representatives of the major companies in the field; and the other of Government officials of the principal producing countries: Canada, South Africa, Australia, and France. There were no American representatives at the meeting: American anti-trust laws would not permit any to attend, and besides, the United States was at this time isolated from the rest of the world market.

One company representative recalls the atmosphere of that first meeting: 'We just came together and wept on each other's shoulders.' But after the tears were shed, there was business to be done. They agreed in principle to divide up the world's markets, giving each producer a share, and to fix a price below which the uranium would not be sold. To organize this, they would set up what was to be called the Uranium Research and Marketing Organization which was a euphemism for a cartel.

The company representatives who assembled in April, and at subsequent meetings, were men for whom uranium is simply a commodity, for buying, selling, speculation and investment, like copper or silver, cotton or wheat. Its end use interests them only insofar as it affects requirements for it. Some of them had trained as geologists and started their careers underground with a pick in their hands, men like Mario de Bastiani of Denison, a veteran of the brawling, pioneer days at Elliot Lake. Others had begun in the office and worked their way up the corporation ladder, like George Albino, a trim, quietly-spoken American and former marine officer who represented the Canadian company Rio Algom, and Tony Gray of the Australian company Pan Continental, a youngish lawyer, who impressed the others during negotiations as having a mind that, as one of them put it, was 'logical but not legalistic'.

Governments were represented heavily. Uranium produced in

France and Francophone Africa is marketed exclusively by a French Government corporation run by civil servants, Uranex. The Canadian Government set up a company to manage and sell off its stockpile, Uranium Canada Ltd., or Ucan, and this was represented at the Paris meeting by officials of the Ministry of Mines and Resources. Nuclear Fuels of South Africa Ltd., Nufcor, the uranium producers' marketing organization, was also represented by a civil servant, John Shilling.

There was some hard bargaining over quotas. Everyone was being assigned quotas that would still leave them with unsold uranium. The Australians said their quotas should be raised because their new mines could produce uranium more cheaply than the others, and sell it more cheaply if the cartel were not established. The Ucan representative warned that if it came to selling cheaply, Ucan had a huge stockpile which was already bought and paid for, and it was quite ready to sell at a price far below the production cost just to get rid of it. Shilling said South Africa had a uranium refining mill standing idle, just waiting for ore to be fed into it. An Australian, John Proud of Peko-Wallsend Ltd., warned that this discord was only benefiting the power companies which buy uranium, because if it went on, he could see the price falling still further, to $4 a pound.

Agreement was reached finally. Canada was to have about a third of the market, to be divided among Canadian companies as they agreed among themselves; South Africa would have 24 per cent; France and Francophone Africa 22 per cent; the new Australian producers 17 per cent; with the old Australian company Mary Kathleen, now owned by RTZ, and Rossing in Namibia sharing the remainder.

A secretariat was set up in Paris, in the CEA office building, run by Andre Petit, a CEA official and the French representative on Euratom. A company participating in the arrangement would pass on any requests for uranium to the secretariat. Petit would then steer it to the producer whose turn it was, and he would offer it at the agreed price, or a higher one if he thought he could get it. A designated runner-up would ask a price 8 cents a pound higher, and others would quote higher prices still. The cartel did not publicize its existence, but people in the industry knew about it, and it was referred to occasionally in the trade press. It was called 'the club'.

The club met again several times, in Johannesburg in the Chamber of Mines building, in London in the RTZ offices, and once in a hotel in Las Palmas in the Canary Islands. When it came to fixing prices, they were not greedy; they did not push prices up very much. At the first Johannesburg meeting, in June 1972, they fixed a minimum price at $5.40 for uranium to be delivered immediately, and $7.50 a pound for uranium to be delivered in six years' time, in 1978; much of the uranium sold is for future delivery. The minimum prices went up at successive meetings, until the last meeting, in June 1974, when the prices were $8.50 and $12.50 respectively. Looking ahead at the time when the American market might open up, the members agreed that their arrangements would not apply to the United States.

There were disagreements from time to time. RTZ was accused of selling uranium to Spain when it was somebody else's turn. Louis Mazel, the RTZ representative at most of the meetings, defended his reputation so fiercely that at one point the others were afraid that he and one of his accusers would come to blows. On another occasion, there was a dispute about whether a commission paid to an agent in a sale was in fact a discount to the buyer which infringed agreed rules.

Governments supported the cartel. The Canadian and Australian governments indicated that they would not give export licences to companies to sell uranium at prices below those set by the club.

Starting near the end of 1973, the market price of uranium, after sliding downwards for a decade, took a sudden upturn. Several events contributed to the increase in demand, some of them far removed from the fields of uranium and nuclear power. In Australia, the Labour Party Government elected the year before suspended all export licences while it pondered the morality of sustaining nuclear power. In America, the AEC announced that it would phase out its embargo on foreign uranium. Then it increased the tails assay, the amount of uranium that could be left in the discarded ore after extraction, which meant that more uranium ore would be required to produce a given quantity of pure uranium. The Arabs' oil embargo in late 1973, and the subsequent quadrupling of oil prices, spurred a new interest in atomic power.

Suddenly, expectations changed. People worried about an

energy shortage, and about supplies of every kind of fuel. They wanted the security of stockpiles of fuel to fill their needs for years ahead. The Canadian company Ucan got an idea of the way the wind was blowing when representatives of the Spanish Government's newly formed atomic power company came to Canada to buy uranium for the reactors the company was planning to build. They thought the Spaniards might want several hundred tons; the visitors said they were looking for 34,000 tons. This was seven times Canada's annual production, and enough for Spain's needs for decades to come. In the end, the Spaniards bought 8,000 tons from several sources. British Nuclear Fuels Ltd., signed a contract with Rio Algom for the purchase of uranium over twenty-five years. The Tennessee Valley Authority, looking ahead to the end of the US embargo on uranium imports, invited bids to supply it with 43,000 tons of uranium, which would be the largest order ever placed by an American utility. Stockpiles shrank, and the price of uranium for immediate delivery went up from $6.50 in June 1973 to $15 in December 1974.

A cartel which aims to prevent the price of uranium from falling below $8.50 a pound is superfluous when people are paying $15 and the price is still climbing. At their meeting in Johannesburg in September 1975, the members of the club wound it up, and went back to competing with one another in the ordinary way. The price of uranium went on climbing, to a peak of $41 a pound in 1976.

* * *

The club was dead. But it was to be far more contentious after its demise than it had been during its lifetime. This was largely because of the trading practises of the Nuclear Reactor Division of the Westinghouse Electric Company. Westinghouse and General Electric are the two giants in the manufacture of electrical appliances, and of the two, GE is the bigger and more profitable. They both entered the business of producing nuclear power plants at the same time; GE specialized in the boiling water reactor, and Westinghouse in the pressurized water reactor. The PWR was the bigger seller, and Westinghouse pulled ahead in this one area. By the 1970s, the competition between the two was intense because both had lost a lot of money on fixed-price reactor sales. Westinghouse's vice-president in charge of its Nuclear Reactor

Division was John W. Simpson. He was sent by Westinghouse to Oak Ridge for a year as a promising young engineer in 1946, and returned an evangelist for nuclear power. A quiet-spoken but earnest man, he used to tell people that nuclear power would not only put Westinghouse ahead of General Electric, but would also enrich humanity and eradicate hunger and want from the world.

In the reactor field, Westinghouse acquired a reputation for an aggressive, even buccaneering style of selling. For instance, when the Canadian CANDU reactor was competing in some markets, Westinghouse reactor salesmen had a presentation kit on the CANDU complete with colour slides designed to show that it was less safe and less reliable than the PWR.

Westinghouse started offering an extra inducement to potential customers: a guarantee of a twenty-year supply of uranium, at a fixed price of about $10 a pound. Other reactor companies admitted that this offer gave Westinghouse a competitive edge. It signed twenty-seven of these contracts, twenty-two with American utilities and the others with Japanese and Swedish companies. It even signed some of these contracts in 1975, when the price of uranium had already started its climb. It refused offers of uranium at $7 a pound. Economists at Westinghouse offices in Pittsburgh decided that the price was cyclical, like the price of copper, and it would soon fall again. When it reached $40 a pound, they stopped waiting for it to fall.

Westinghouse faced the prospect of buying some 80 million pounds of uranium over twenty years at $41 a pound, and selling it to its customers at $10 a pound. The loss would be around $2½ billion. Westinghouse told its customers it could not fulfil its contracts. It took refuge in a clause in the law of contracts which exempts a party from commitments if it is prevented from carrying them out in certain circumstances by an external factor 'that could not reasonably have been anticipated.' According to the trade paper *Nucleonics Week*, the nuclear power industry was unsympathetic to Westinghouse's plea: the industry took the view, the paper said, that Westinghouse had gambled and lost, and must now pay up.

This began a series of law suits and counter-suits involving many companies and several governments, which are continuing still. To trace them in any but the broadest outline, would carry us far beyond the scope of this book. Suffice to say that lawyers' children yet unborn will have their higher education paid for out of the fees for law suits which are a part of this imbroglio.

The utilities – the American power companies – sued Westinghouse, starting with the Kansas Gas and Electric Corporation and the Virginia Electric and Power Company. The hard-pressed Westinghouse lawyers found a way to go over to the offensive: they invoked the existence of the club, a price-fixing arrangement that offended against the ethics of American business.

In this, they received help from an unexpected quarter. In Melbourne, Australia, a brown paper parcel was left by a person or persons unknown on the doorstep of the local office of Friends of the Earth, the worldwide organization that campaigns against nuclear power, among other things. It contained hundreds of letters and confidential memoranda stolen from the Mary Kathleen Uranium, the only company in Australia at that point actually producing uranium. Together, these letters and memoranda gave an account of the club, its meetings and its decisions. FOE did not want to be accused of keeping stolen property, so they gave the documents back to the Mary Kathleen company, but first they made photostat copies. They sent some of the more revealing of these to newspapers. They also made them available to Westinghouse lawyers, at the lawyers' request.

Westinghouse sued twenty-nine mining companies, American and foreign, in a Federal court in Chicago, alleging that a criminal conspiracy to keep up the price of uranium prevented it from fulfilling its contracts on commercially acceptable terms. It named American producers among them, including most of the biggest; Kerr-McGee, Anaconda, the Getty Oil Company, and the United Nuclear Corporation. It charged that these companies, even though they were not members of the cartel, co-operated with it to raise prices, that they exchanged pricing and marketing information with the cartel, and that they either refused to sell uranium to Westinghouse or else offered discriminatory and unacceptable terms. It demanded triple damages, which is normal where criminal conspiracy is alleged: 6 billion dollars. As the legal arguments and counter-arguments mounted in complexity, the defendants assembled two million documents, and they found it profitable to join together to buy a computer system for a million dollars to handle them all.

The mining companies said that the cartel had not pushed their prices up, and that the Westinghouse argument was simply a weak alibi for the corporation's miscalculation. This view was supported by the judgement of the Nuclear Exchange Corporation, Nuexco, the uranium brokerage firm in Fresno, California; Nuexco handles a substantial part of the world's uranium trade, and is regarded as an authoritative source of market information. In a report it published on the rise in uranium prices since 1974, it referred to the club, and said:

> It is Nuexco's judgement that the activities of this gathering . . . had little or no effect on domestic uranium prices. . . . The prices reportedly established by the 'cartel' at its periodic gatherings were, by the time such prices were implemented, below those then prevailing in the domestic market place. The 'cartel' was, in fact, in a position of trying to catch up with the domestic market, as opposed to leading it.

Eight of the utilities' lawsuits against Westinghouse were consolidated, and brought before Judge Robert Mehrige Jr in a Richmond, Virginia court. Westinghouse said it needed some documents from RTZ on the cartel as part of its defence. RTZ declined to hand over the papers. The Richmond court asked the British high court to compel RTZ to do so, and the senior master in the high court acceded to this request. RTZ appealed, and the British appeal court said RTZ need not hand over documents, but must give oral evidence. So Judge Mehrige flew to London, and held a hearing in a room on the fourth floor of the US Embassy overlooking Grosvenor Square, and seven RTZ executives were summoned to give evidence.

There, Judge Mehrige was treated to the droll spectacle of some of the most distinguished figures of the British business world, including the Chairman of RTZ, Sir Mark Turner, none of whom were even suspected of breaking their country's laws, going to the witness stand and 'taking the fifth', that is, refusing to testify, claiming the right provided by the fifth amendment to the US constitution not to give evidence that might incriminate them.

By now, the US Government was involved. With the rise of consumerism in the 1970s, and the elevation of Ralph Nader to the status of a national hero, there was an aggressive attitude to enforcing anti-trust legislation, particularly under the Carter

Administration. Some US courts carried their pursuit of cartels beyond American territorial limits.

Anti-trust laws have a significance in America that they do not have in other countries. In most countries, a cartel is judged pragmatically. But in America, free competition in commerce is seen as an ethical value, indeed, a part of the moral foundations of the Republic. A cartel, because it violates this, is deemed to be evil in itself. Monopoly is the point at which business becomes unAmerican. When an American executive of Gulf Minerals, the Canadian subsidiary of Gulf Oil, was questioned by a Congressional subcommittee about his company's participation in the uranium cartel, he was asked at one point, 'How does all this square with your personal moral code?' He admitted that it posed a problem of conscience for him. It is difficult to imagine anyone other than an American feeling participation in a cartel to be so fundamental a moral question.

There was pressure on the US Government to look into the uranium cartel from several sources: Westinghouse, some congressmen, and others who were aroused by an article in *Forbes*, the widely-read business magazine, warning against 'a uranium OPEC'. The US Justice Department started an investigation, and grand jury hearings.

The Justice Department looked forward to hearing the RTZ evidence in the Westinghouse case, since this would be a firsthand account of how the cartel fixed prices. After Sir Mark Turner and the other RTZ chiefs took the fifth, the Justice Department immediately made a counter-move: it told the court that the RTZ men could have immunity from prosecution. This meant that they could no longer say their evidence might incriminate them. Now RTZ appealed to the House of Lords, asking it to reverse the appeal court's ruling that it must testify. Lord Dilhorne and the four other law lords concerned granted RTZ's case. The reason, they said, was that the evidence was required primarily not for the Westinghouse case, but for a grand jury investigation into activities of companies that were not under the jurisdiction of the United States, and that to drag in a British company would be an infringement of British sovereignty.

Several other governments besides the British felt that their sovereignty was in danger of being infringed. They did not regard decisions reached by British, French, Canadian, Australian and South African companies, in meetings in Paris, Johannesburg and London, as a fit subject for investigation by the United States Government. It was not only the US Justice Department that was carrying out an investigation. A subcommittee of the House of Representatives Commerce Committee held a series of hearings into the cartel, at which the principal witnesses were executives of Gulf Oil. The Canadian Government vented its annoyance at all this by releasing letters showing that it had told the US Atomic Energy Commission from the beginning that producers were getting together to fix prices, implying that shock and outrage in Washington were not to be taken too seriously. The Canadian Government also passed legislation to make it illegal to give a foreigner official information on uranium or atomic energy matters, so that if any foreign court tried to compel a Canadian to testify about Canadian activities, he could plead that he was prevented by his country's own laws from doing so.

The US Attorney General, Griffin Bell, visited Ottawa to discuss the question of the jurisdiction of American courts in these matters, and Canadian officials put their concerns to him. Immediately after his return, he gave an address to a meeting of the American Bar Association in Chicago that showed that he was unrepentant, and determined to press on.

> We are scrupulous in not reaching beyond our authority, . . . But our law enforcement obligation does not allow us to look the other way when an anti-trust investigation involves foreign nationals. . . . We are obliged to do all we reasonably can to prosecute foreign cartels which have the purpose and effect of causing significant economic harm in the United States in violation of our anti-trust laws.

Canadian, British and Australian officials consulted on this speech, and found it hawkish.

Australia followed Canada in passing a law forbidding the handing over of information on uranium to foreigners, and then another permitting the Government to quash a foreign anti-trust judgement. South Africa, a less open society, had a law restricting information on its statute books already.

Britain followed with the Protection of Trading Interests Bill, which was designed to block the jurisdiction of American courts, not only in the uranium case, but in anti-trust suits in general, and also in civil suits such as the Westinghouse one against RTZ in which triple damages were claimed. The Trade Secretary, John Nott, introduced the bill with an unusually strong statement which left no doubt about its target. Its purpose, he said, was to 'strengthen our defences against US practices which are not only widely regarded as unacceptable internationally, but are having a most damaging effect on the commercial activities of British companies.'

The grand jury investigations and the Congressional hearings circled around the two questions of whether some Americans had to pay more for uranium because the cartel was in operation, and whether the American company Gulf Oil violated anti-trust laws because of the participation of its Canadian subsidary. In the end, Gulf Oil was found guilty of a misdemeanour, and fined a trifling $40,000.

Meanwhile, the Tennessee Valley Authority, which had ordered 17 million pounds of uranium from Rio Algom, sought to cancel its purchase on the ground that the cartel had pushed up the price. The TVA found it suspicious that it had asked fifty-three American and foreign companies to quote a price for supplying a large quantity, and received only three replies. Rio Algom sued the TVA for breach of contract, for $1.2 billion.

Governments were involved in the Westinghouse suit against the uranium producers because the non-American companies which were sued claimed that the court had no jurisdiction over them, and their governments backed them up. The British, French, Canadian and Australian Governments filed submissions in the Chicago court. The US Government got involved because John Shenfield, then the head of the Justice Department's anti-trust division, wrote to the court suggesting that no sanctions should be imposed unless the defendants had acted in bad faith. This annoyed some cartel-hunting congressmen, who felt that the Government was interfering improperly to soften the court's hand, and truckling to foreign interests.

Eventually, in 1981, all the cases were settled out of court, some of them individually and some in groups. Westinghouse paid out a total of more than $450 million to the utilities that had brought law suits against it, and agreed to supply 5 million pounds of uranium, a small fraction of the 80 million pounds that was originally contracted for. In its case against the major uranium producers, Westinghouse accepted settlements that brought it something over $100 million. The settlements also involved the transfer of some uranium contracts and other complicated financial arrangements.

Meanwhile, the price of uranium has gone into a depressive phase. Starting in early 1979, it slid downwards, from $41 a pound to less than $30. Taking inflation into account, this means a drop of more than 50 per cent. It looks like being a buyers' market again.

Chapter 9
TURNABOUT

It is difficult to practise export restraint in an atmosphere of competitive free enterprise. KWU won the lucrative order to supply reactors to Brazil against foreign competition because it was prepared to sell enrichment and reprocessing plants as well. What was needed was an agreement among exporters on what items should be removed from commercial competition, and this was reached at a series of meetings in London.

The initiative came from a Canadian, Michel Dupuy, an Under-Secretary at the Ministry for External Affairs, who had gone with Ivan Rowe to India on the unsuccessful mission to restore Canada's nuclear power relations. Dupuy decided that other countries besides Canada should change their ways in the wake of the Indian explosion, and that the restrictions laid down by the NPT were not sufficient. Dupuy's idea was accepted by the Canadian Government, and then he tried it out in tentative conversations in several capitals. He knew it would have to be accepted by the Americans if it was to get off the ground and he worked on the State Department first. Dr Kissinger, who was then Secretary of State, gave it his backing.

The first response from others was not encouraging. In Britain, the Foreign Office and the nuclear establishment tended to feel that the spread of nuclear power had little to do with the spread of nuclear weapons, and any restrictions beyond the NPT would be an irrelevant handicap on trade. But David Owen, the new Foreign Secretary, made nuclear proliferation his personal concern, and decided that more restrictions might be needed.

France was an important nuclear exporter, as the projected Pakistan deal demonstrated. France had disdained to take part in any international arms control discussions since de Gaulle set it on its course of strict and even doctrinaire nationalism in 1958. Its chair at the UN Disarmament Committee is empty. It had shown no interest in joining the NPT. The Director of the US Arms Control and Disarmament Agency, Fred Iklé, a Swiss-born and French-speaking academic, broached the idea of an agreement to restrict exports with French officials on a visit to Paris. They objected that such an agreement among the industrialized countries to restrict exports would be unfair to the Third World. He found this hypocritical and went away angry. But Giscard's Guadeloupe conversion to the cause of international anti-proliferation measures produced a change, and France agreed to take part in a meeting to discuss restraints.

Russia also responded positively, and this was welcomed in the West. Russia was exporting nuclear reactors to its allies, and small research reactors and other materials to a few Third World countries. These are sold under stringent conditions that surpass the NPT requirements. There is no question of a customer asking Soviet permission to reprocess the used fuel from a Soviet reactor; all the used fuel must be returned to Russia.

The major suppliers met in London in April 1975, and again later the same year, and the group came to be called the London Suppliers' Group. The meetings were secret, and were announced only after they had taken place. This was at the insistence of the French Government, which was already being accused by die-hard Gaullists at home of insufficiently staunch defence of French sovereignty in NATO and the EEC. Representatives came from Britain, Canada, France, Germany, Japan, the Soviet Union and the United States. These were the new kinds of men who were appearing at international discussions, men from foreign ministries with a background in diplomacy who had taken up nuclear power issues as a speciality, and had studied the uses of uranium and plutonium and had ideas on the subject.

These meetings laid down rules which have been observed since in the nuclear trade. The most important is an agreement by the seven not to export any more enrichment or reprocessing plants. Germany refused to cancel its contract with Brazil, but agreed not to do the same again. They also drew up a list of items that they would not export.

Other agreements plug up a few loopholes that had appeared in the NPT. The seven countries agreed that they would sell nuclear material only when the recipient country agreed to international safeguards over it with no provision for termination by withdrawal from the treaty. They will insist on control over what is called 'replication', that is, a receiving country copying the design of a facility that has been sold in order to build another.

Most of the seven wanted to go further, and limit exports to countries that accept 'full-scope safeguards', which means opening all their nuclear facilities to international inspection. But France and Germany refused to accept this, and have continued to do so. Other participants in the discussions believe that France would have been willing to agree to it, but Germany held out, and the French Government could not afford to appear less nationalist than the German.

These agreements are not treaties. They consist of unilateral statements made jointly by the signatory countries, and lodged with the International Atomic Energy Agency.

Later, other countries joined the London Suppliers' Group, for two more secret meetings in 1976. These are smaller countries with high-technology industries that export some components used in nuclear power: Belgium, Czechoslovakia, East Germany, Holland, Italy, Poland, Sweden and Switzerland. These have all accepted the same restrictions. Among nuclear suppliers, only two remain outside the group, South Africa and India. Either or both of these could turn into rogue elephants so far as nuclear trade is concerned.

These agreements have tightened the guidelines on exports usefully, but at a price. The new conditions are being imposed on consumer nations by the suppliers acting together. This has produced resentment in some other countries, which see a club of rich, industrialized nations dictating the terms on which they will have nuclear power. This resentment is felt all the more deeply by countries which signed the NPT and therefore voluntarily accepted restrictions.

The Director of Argentina's National Atomic Energy Commission, Castro Madero, gave expression to this resentment, and to a widespread suspicion about the motives of the big exporters, when he addressed a nuclear power committee of non-aligned nations at its first meeting in July 1980. 'The London Club's attempts to protect its economic interests by using the non-proliferation issue have almost paralyzed international trade in this field,' he said, exaggerating the position somewhat. 'Moral principles that we all share should not be preached to protect the political and economic interests of the few.'

The sentiments were directed at a group of thirteen industrialized countries. Since then, similar sentiments have been directed by others within this group against the United States and Canada. For it is the nature of the nuclear trade that exporting countries are also importing countries.

* * *

The production of books, articles in specialized journals, and papers and reports on public policy issues has become a considerable industry in the United States. Philanthropic foundations, Government-funded research institutes, think tanks, and university departments fund scholars, individually and in groups, to write these. Usually, these are read only by others in the academic world, even when they are paid for by a Government department. It might be said for them that they sometimes stimulate thinking, and keep a number of young people off the streets.

Very occasionally, however, one of these produces a direct effect on Government policy. In 1975 and 1976, two such studies altered the US Administration's policies on nuclear power, and set off a worldwide argument that is still going on. One was produced by Pan Heuristics, a California research institute that does work for government agencies. The other was produced by a group of academics assembled by the Mitre Corporation, a Boston research organization, at the behest of the Ford Foundation.

The first study was carried out on the initiative of the Director of the Arms Control and Disarmament Agency, Fred Iklé. Dr Iklé himself worked at the Rand Corporation, the grandfather of the Government-aided think tanks. He is a thoughtful man with an academic background in political science, the author of, among other things, a book on international negotiations and a widely discussed article on the immorality of nuclear deterrence. He worried now about the connection between the spread of nuclear power and the spread of nuclear weapons. ACDA, was playing little part in the most important arms control activity at this time, SALT, because Dr Kissinger had taken charge of this himself. Proliferation, Iklé felt, was one area in which ACDA should get involved.

He commissioned a study on the subject from Pan Heuristics, a think tank founded and headed by Albert Wohlstetter, another Rand alumnus.

Wohlstetter was one of the first 'defence intellectuals' of the 1950s and '60s, those professional thinkers who take a long, cool, and above all, analytical look at some of the vital issues of our times in order to advise a government department or agency, and he has probably had more influence on events than any other. At Rand in the 1950s, he produced a report for the US Air Force in which he created the conceptual framework of nuclear deterrence, drawing for the first time the crucial distinction between a first and second strike force. Now he combines his work at Pan Heuristics with the post of University Professor at the University of Chicago, which means that he is not attached to any one department but can lecture on any subject he likes, and he has homes in both Chicago and Southern California. He is a stocky, trim figure with a small white beard, soft-voiced and courteous. He has rigorous professional standards and reaches conclusions painstakingly, and then presses them forcefully. He has refused offers of Government posts because he does not want to deal with day-to-day problems, but he knows the ropes in Washington, and knows which ones to pull.

For this study, Wohlstetter gathered a team of twelve assistants, including his wife Roberta, a historian of distinction, and examined the economics of nuclear power and nuclear exports policy. He and his wife went to India and South Korea and talked to officials there, and to many people in the American nuclear power industry and in the Government. He was surprised at some of the vagueness and ambiguities in American export policies, on reprocessing, for instance, and on highly enriched uranium for research reactors.

They concluded in their report that, contrary to the widely accepted view in the nuclear power industry, using plutonium as fuel in thermal reactors is unnecessary and uneconomic. They also said that it would probably prove uneconomic to use it in breeder reactors and certainly that there was no economic justification for preparing to do so now. They looked at the cost of reprocessing, which was continually rising, and compared recycling via breeders with the cost of buying more uranium instead. They recommended strongly against any early commitment to a breeder. The Pan Heuristics report poured cold water on many of the assumptions about breeder reactors that had been accepted widely for years. Thus:

> It is uncertain when, if ever, the breeder will become viable commercially. . . . Even if it should become economic, still more uncertain are the dates at which it would become so, and the realistic rate and extent at which it might grow in commercial significance without risk of losing enormous investment. The immense uncertainties are confirmed by the poor record of the United States and the United Kingdom in the field of the spread of civilian and military nuclear energy. They all suggest the importance of strategies of sequential decision which avoid a premature commitment, and proceed to commitment as uncertainties are progressively resolved.

The report was detailed and closely argued, with calculations of the costs of various stages of the nuclear fuel cycle, alternative ways of using nuclear power, and, with regard to Third World countries, alternative forms of power.

The economic argument against using plutonium was one part of the conclusion, but this was not the most important feature, nor did this inject into the report the passion that sometimes strains at the language of cool analysis. This was the other side of the argument: that the use of plutonium increases the likelihood that nuclear weapons will spread.

The report links the two in the case of India. It says India's investment in a reprocessing plant to extract plutonium for use in a breeder reactor has turned out to be premature and used scarce resources which could have advanced economic growth. On the other hand, possession of the reprocessing plant reduced both the time and money required for India to build its own atomic explosive and hence made it easier to decide to do so. The tone of alarm was sounded in the title under which the report was first sent to ACDA and others in the Government: *Moving Towards Life in a Nuclear Armed Crowd.*

It impressed Iklé, and stirred him to action. He was already worried about proliferation. He was convinced by the reasoning of the Pan Heuristics report, and he wanted the Administration to do what it could to prevent anyone from rushing into reprocessing used fuel and breeder reactors, and therefore into a heavy investment in plutonium. He got support from the Secretary for

Defense, Donald Rumsfeld, and from the Bureau of the Budget, where officials were very willing to be persuaded that the hundreds of millions of dollars that were being paid in subsidies for the construction of a reprocessing plant were unnecessary. The Energy Research and Development Agency, one of the two successors of the AEC, was against him; they did not want the option of an early breeder reactor programme closed. The State Department was doubtful, expressing concern at the response of America's allies to an attempt to tell them what their nuclear power policies should be (correctly as it turned out).

Fred Iklé wrote of the Pan Heuristics report:

> Rarely has scholarly research been so influential for changing government policy. This study, far more than any others on this topic, revolutionized the thinking in the United States (and in other countries as well) leading the way to the radical new departure in US non-proliferation policy that took place during the Ford Administration.*

Another study also influenced informed and political opinion: what came to be known as the Ford/Mitre Study. This was carried out on the initiative of the Director of the Ford Foundation, McGeorge Bundy, former National Security Advisor to Presidents Kennedy and Johnson.

Bundy had been impressed by a report on energy done for the Ford Foundation by David Freeman, and he decided that the Ford Foundation should do some work on nuclear power, about which people were asking new questions, or at any rate, asking old questions more audibly and more urgently. He tried to bring opponents and supporters of nuclear power together for a series of discussions, but concluded after several approaches that the opponents were not really interested. So he went to Spurgeon Keeney, the Vice-President and Director of the Mitre Corporation, which does studies for government and industry, and asked him to gather some people around him to produce a report. He felt that Keeney was uncommitted on the subject and wanted others who

---

* In his preface to *Swords from Plowshares: the Military Potential of Civilian Nuclear Energy*, which is essentially the Pan Heuristics report published in book form.

were also. The twenty-one people who made up the team are the kind of academics who have had some contact with government operations, either in official government positions or as outside advisers. Most of them knew each other. (A prominent member was George Rathjens, then a Professor of Political Science at MIT, the past author of studies on nuclear deterrence and arms control for the Pentagon. Rathjens and Keeney used to play in the same Sunday morning touch football game in Washington; Keeney, a smallish, wiry man was the quarter-back and a good passer, and the tall, rangy Rathjens was usually his receiver.)

They hired a small staff, they consulted a number of people in government and in the nuclear power industry, they held meetings during 1976, including a ten-day one in the summer, and published their report in January 1977, calling it *Nuclear Power Issues and Choices.* It was longer than the Pan Heuristics report, running to 418 pages and it became much more widely known, partly because it was published as a book right away, and Bundy ensured that it received media coverage. In the international controversies that developed later over what became the US Government's policy on nuclear exports, this policy was identified widely abroad with the Ford/Mitre report, and was sometimes referred to in conversation simply as the Ford/Mitre policy.

Its conclusions were similar to those of the Pan Heuristics report. It found that nuclear power has an economic edge over coal in producing electricity, but not a large or significant one, and found that in terms of damage to the environment and other social costs, there was no way to measure nuclear power against coal power. However, like the Pan Heuristics report, it said that economic competitiveness was not the crucial question:

> By far the most serious danger associated with nuclear power is that it provides additional countries with a path for access to equipment, materials and technology necessary for the manufacture of nuclear weapons. . . . We believe the consequences of the proliferation of nuclear weapons are so serious compared to the limited economic benefits of nuclear energy that we would be prepared to recommend stopping nuclear power in the United States if we thought this would prevent further proliferation.
>
> However, there are direct routes to nuclear weapons in the absence of nuclear power, and the future of nuclear power is not under the

> unilateral control of the United States. Most advanced countries are now actively developing and utilizing nuclear power, while less developed countries count it among their expectations. In fact, abandonment of nuclear power by the United States would lose influence over the nature of nuclear power development abroad. With continued nuclear power development, however, the US Government must give greater weight to the proliferation problems in its decisions on nuclear matters and its relations with other nations.

It suggested that the United States should give a lead in dropping plans to reprocess used fuel for breeder reactors:

> Our net conclusion is that reprocessing and recycle are not essential to nuclear power, at least during the remainder of this century. In addition, there are potentially large social costs, including proliferation and theft risks, in proceeding. A US decision to proceed despite disincentives would induce other countries to follow suit, and undermine efforts to restrain proliferation.

* * *

By the summer of 1976, the alarm bells that had sounded with the disclosure of the projected German-Brazillian deal and the French plan to sell a reprocessing plant to Pakistan were being heard in many quarters. A few members of both houses of Congress took up the cause, and when the Senate passed the 1976 Foreign Assistance Bill, an amendment first proposed by Senator Stuart Symington required the Administration to cut off all aid except food to any country which buys or builds a reprocessing plant without safeguards.

It is important that at this time, the link with nuclear weapons was not the only charge that was being made against nuclear power. The atmosphere was very different from that of a few years earlier, in which some enthusiasts for nuclear power carried along with them a trusting public. Now some people were becoming worried about the safety of nuclear power and doubtful about the sense of social responsibility of the men who were managing it. It was easier to criticize nuclear power now; it did not stand in universal good esteem, without a blemish on its character.

When Governor Jimmy Carter of Georgia sought the Democratic nomination for the Presidency, he gathered around him, as most Democratic contenders do, some advisers from the campuses of Cambridge, Massachusetts. Two from Harvard were Abram Chayes and Joseph S. Nye Jr., both of them members of the Ford/Mitre Group. The ideas that went into the report were already being formulated, and were the basis of their advice to Carter on nuclear policy. Also, Spurgeon Keeney had circulated copies of the Pan Heuristics Report to all the members of the group, and this was read carefully. Wohlstetter wrote a paper on non-proliferation for the Carter policy group.

Jimmy Carter was the first national leader anywhere to have been involved in nuclear technology. As a career naval officer newly out of Annapolis, he became one of Admiral Rickover's young men, and trained as a nuclear engineer to work on nuclear-powered submarines. But he emerges in conversation as very cautious on the subject, aware through his training of the dangers as much as the potential benefits. He listened to Chayes and Nye, and responded to their concern.

In the Spring of 1976, Carter was accused by some of his Democratic rivals of presenting the American public with little more than a bland, toothsome smile and some pleasing sentiments about decent government, and not addressing himself to specific issues. So his staff let the political reporters know that he would set about doing so, and he started with a speech on nuclear power, in New York City in May.

Carter deplored the French and German sales of nuclear reprocessing plants, and said President Ford could have headed off these sales, 'as he should have done many months ago', if he had taken action at the highest level.

When he turned to the wider issues of nuclear power, he sounded the tones of the doubters rather than the enthusiasts: 'US dependence on nuclear power should be kept at the minimum necessary to meet our needs. We should apply much stronger safety standards as we regulate its use. And we must be honest with our people concerning its problems and dangers.' He dwelt, not on the benefits of nuclear power, but on its hazards: possible accidents, radioactive waste, the terrorist threat, and proliferation. He urged development of alternatives to nuclear energy. This kind of thing had not been heard from the top before, in America or anywhere else.

He took up the theme of proliferation again the following

month, when he addressed a United Nations' meeting. This time, he suggested that reprocessing facilities should be taken out of national ownership, and that it should be done at multi-national reprocessing centres. Wohlstetter, who was aware of his own input into the Carter campaign, was present and heard the speech. He tackled Cyrus Vance, then Carter's principal foreign policy adviser, in the lobby, and told him Carter still had not got it right, because a multi-national reprocessing centre would still keep separated plutonium, and send it to countries that do not now have nuclear weapons.

When the election campaign was in full swing, Fred Iklé saw in Carter's speeches a way to advance the Pan Heuristics–Ford/Mitre ideas. He went to President Ford in the White House and urged this view on him. He pointed out that Carter was playing up nuclear power and proliferation as campaign issues and blaming Ford for not having done more. 'But you've got a fine track record in this area,' he said. 'You persuaded Giscard on the dangers of proliferation. You were instrumental in getting the London Suppliers' Group going. Why let Carter make the running now?'

So, on 28 October 1976, in the middle of an election campaign, President Ford set American nuclear power policy on a new course.

Setting the tone for the policy statement, he said: 'Developing the enormous benefits of nuclear energy while simultaneously developing the means to prevent proliferation is one of the major challenges facing all nations of the world today. The standards we apply in judging most domestic and international activities are not sufficiently rigorous to deal with this extraordinarily complex problem.'

He went on: 'I have concluded that the reprocessing and recycling of plutonium should not proceed unless there is sound reason to conclude that the world community can effectively overcome the associated risks of proliferation. I believe that avoidance of proliferation must take precedence over economic interests.'

He took a more positive attitude to nuclear power than Jimmy Carter, however: 'I have also concluded that the United States and other nations can and should increase their use of nuclear power for peaceful purposes even if reprocessing and recycling of plutonium are found to be unacceptable.'

If he was to recommend a new attitude to the nuclear fuel cycle for the world, he would have to order the same change in the United States, and he did: 'I have decided that the United States should no longer regard reprocessing of used nuclear fuel to produce plutonium as a necessary and inevitable step in the nuclear fuel cycle, and that we should pursue reprocessing and recycling in the future only if they are found to be consistent with our international objectives.'

He dwelt on America's role as an exporter, and on the need for international as well as merely national controls. Thus: 'Given the choice between economic benefits and progress towards our non-proliferation goals, we have given, and will continue to give, priority to non-proliferation. But there should be no incompatibility between non-proliferation and assisting other nations in enjoying the benefits of peaceful nuclear power if all supplier countries pursue common export policies.'

He ordered the Energy Research and Development Agency to change its policies accordingly, and to stop Government funding for construction of the Allied Chemical Company reprocessing plant at Barnwell, South Carolina, which meant halting the project. No longer was plutonium, in the US Government's eyes, a potentially useful fuel, to be assigned a cash value. Now it was a highly dangerous substance, to be kept out of circulation.

The priority granted to non-proliferation concerns over the economic benefits of nuclear power strike a very different note from the speeches of the atoms-for-peace era and the text of the NPT, where there is no suggestion of any conflict between them. If one wants to take two major turning points in the international use of nuclear power, they are President Eisenhower's speech to the United Nations in which he uncaged nuclear power with the atoms-for-peace programme, and President Ford's statement in October 1976, in which he tried to cage it again.

Chapter 10
THE URANIUM CYCLE

Carroll Wilson, the first General Manager of the AEC, wrote an article in 1979, long after he left the post, looking back on his years with the AEC and the development of nuclear power, and he picked out what appeared to him in retrospect to be a fundamental omission.

> No group seemed to have looked at the total fuel cycle as an inter-dependent system. No one had enough clout to set priorities and to argue persuasively that if some critical part of the system were missing, perhaps the whole system would come to a grinding halt. . . . The most striking fact in this retrospective look at our nuclear energy programme has been the lack of awareness that the whole inter-dependent system must be satisfactory.*

By the mid-1970s, this awareness had arrived, and people were looking very closely at the nuclear fuel cycle. This, rather than reactor engineering, is at the heart of most of the big questions concerning nuclear power today. The treatment of uranium is at issue, all the way from its extraction from the ground to its disposal after use in a reactor as radioactive waste.

Today, uranium is extracted from the ground mostly in North America, Southern Africa and Australia. Eighty-five per cent of the uranium mined outside the Communist Bloc comes from these three areas. The United States is the world's biggest producer, mining something over 15,000 tons a year, but it does not sell any abroad. Canada follows closely in output, with 8,000 tons a year, and is the biggest exporter; Australia is producing little but is opening some mines that were closed down during the uranium depression, and has large reserves. South Africa produces more than 5,000 tons a year, but this comes from quartz-gold conglomerates, and the uranium content of those conglomerates is

---

* *Bulletin of the Atomic Scientists*, March 1979

small, so that it is profitable to extract it only because gold is being mined as well.

The pitchblende deposits that were the first sources of uranium, characteristically concentrated but limited in extent, have been mined out. The Shinkalobwe mine was closed down in 1960, and Joachimsthal in Czechoslovakia and Great Bear Lake in Canada soon afterwards. The big veins in the American West and Canada are still being mined, though in most of the ones that have been worked for some years, the grade is getting lower, around 0.1 per cent. More deposits have been found in the west of Canada, and at Cluff Lake, in northern Saskatchewan, the richest deposit ever found, with an ore that is an average of 7 per cent uranium and goes up to 40 per cent in some spots. This is so high that mining presents unusual radiation problems.

Uranium from Namibia, which supplies nearly half the fuel for Britain's atomic power stations, presents a different kind of problem. It is mined by Rossing Uranium Ltd., a consortium in which RTZ has the controlling share. The problem springs from Namibia's disputed international status. The UN General Assembly denies South Africa's right to govern Namibia, and says it has no right to sanction the export of Namibia's mineral resources. SWAPO, the largest single Namibian nationalist movement, and SWAPO's Third World allies say that the export of the Namibian uranium to Britain is therefore illegal. A SWAPO representative, Hadino Hishongwa, told an international conference in 1979 that SWAPO would ban uranium production when it comes to power, because 'uranium is an evil, a threat to Namibia and the entire world'. There have been reports of plans by SWAPO sympathizers to hijack a planeload of Nambian uranium as it flies over other African countries in the name of the United Nations resolution.

Niger and Gabon produce some uranium, and France produces enough for about half its needs. Small amounts of uranium are mined in a number of other countries, Portugal, Spain and Germany among them. The Soviet Union and other Communist Bloc countries maintain strict secrecy about uranium. There are known to be mines in Czechoslovakia, Hungary and East Germany, and there are probably more productive mines in Siberia. Western intelligence agencies believe that the Soviet Union amassed an enormous stockpile of uranium from Eastern Europe in the 1950s and early '60s, and still has some of this.

Uranium mining creates a waste problem in the tailings, the leftover ore from which the uranium has been extracted. This is one more of those problems connected with nuclear power that has crept up stealthily. It was always there, but only recognized as a problem when the tailings piled up. These tailings are radioactive; the extraction of uranium removes only about 20 per cent of the radioactivity, most of which comes from radon, one of the decay products of uranium. At Cluff Lake, where the radon content is phenomenally high, the tailings must be put in a concrete vault and covered with sand. Generally, however, the radiation is very low, but long-lasting. One anti-nuclear group in America challenged an NRC report on the environmental impact of reactors by complaining that it includes the effects of radioactivity in the tailings only for the normal working life of the reactor, which is twenty-six years, instead of for 100,000 years, which it says is the approximate period over which significant radiation is emitted.

Furthermore, tailings do not usually stay in one place, but are shifted about by wind and water. Some 25 million tons of tailings have accumulated in America. Congress passed the Uranium Mill Tailing Control Act in 1978 authorizing three Federal agencies to clean up the tailings, but it did not say how they should do it.

Though uranium mines are located in many different terrains, and though some mines are open cast while some are underground, working conditions in them are fairly similar, because all are in hard rock, nearly all these days are big mines, and in all of them, the substance requires the same treatment.

The mine at Quirke Lake, in Northern Ontario, a part of the Elliot Lake vein, is typical of an underground mine. Seven hundred and fifty men work in a network of tunnels 2,500 feet underground, and they bring up 7,000 tons of ore a day. Uranium mining is not a labour-intensive industry.

The working environment is very different from that of a coal mine, where the miner has to work in the dark in a narrow, cramped tunnel, sometimes so low that he cannot stand upright. The tunnels here are as wide as a four-lane highway, and high enough so that even a tall man in a top hat could stand upright, and they are lit by a string of electric lamps. Ventilating pipes keep a constant breeze blowing, in order to disperse the radon gas. The miners ride through the tunnels in Land Rovers with specially reinforced springs. Bumping along the rocky floor in one of these cars, one can drive for miles without seeing another soul, except perhaps the occupants of another car going in the opposite direction, and one can easily become totally lost by taking the wrong turning at a junction of tunnels.

A constant check is kept on the miners' exposure to radioactivity. Men carrying radiation meters that are about the size of portable cassette recorders check the radiation in the rock face by touching it with an instrument like a stethoscope. It takes only a day's training to learn to use one of these. The number of hours that a miner spends in each section is logged, and he receives a statement of his total exposure with his pay check. The miners are limited to a dose far less than the internationally accepted minimum. This is standard practice in all mines, and in all mines managers sometimes find that some men try to falsify their exposure records, so that they can work more overtime even if it means brushing the safety margins.

The senior men in the mines have been through mining schools and colleges, and have studied metal, or hard rock, mining as opposed to soft rock, or coal mining. Many of them have worked in different countries and several kinds of metal mines. When a number of them from different mines meet, much of their talk is on the lines of 'Whatever happened to old so-and-so?' who went on a two-year contract to the copper belt in Zambia, or to the tin mines in Bolivia. To these men, uranium is a metal to be extracted from ore, like any other. They are concerned with the form it takes in the ore. They have no special interest in nuclear power, and they feel no personal involvement with it, any more than, if they mined silver, they would feel involved with monetary policy.

The uranium deposit at Quirke Lake is a quartz conglomerate, so the walls of the caverns are flinty grey, with patches of white quartz a few inches wide dotted everywhere, surrounded by glittering, silvery pyrite. The uranium is in the pyrite, in specks so small that it is almost invisible. These tiny specks can only be extracted by a very refined sifting process, but it starts with a very crude instrument, dynamite.

Miners carry out the first stage of the blasting with a machine called a jumbo. This is a vehicle with three long drills on top sloping upwards, so that it looks like a mobile rocket-launcher. These drills operate by air pressure, making a horrendous din; the drilling team bores sixty holes 11 inches deep in the rock in a pre-set pattern. The others come with other machines and push ammonium nitrate and then dynamite rods into the holes, and then a detonator, which trails coloured wires behind it. This is going on at rock walls all over the mine. The dynamite rods are all detonated at once when the mine has been evacuated at the end of a shift, so the explosions echo through empty tunnels.

There then ensues the long process of reducing the pulverized rocks left by the explosions to stones, and then pebbles, and then grit, and then powder, so that the chemists can go to work on it. Much of this takes place in a crushing plant and mill that are next to the mine. The rubble is bulldozed on to a conveyor belt that takes it to the mill, where it is pounded by a pile driver and a series of crushing machines.

The result is mixed with water until it is the consistency of mud. This mud is subjected to a series of chemical treatments of which the end product is uranium sulphate, contained in a huge vat 20 feet across, a greenish yellow liquid, steaming fiercely, an awesome sight with an infernal – literally – smell. Then there is more chemical treatment, and at the end the solution is poured along a half-pipe into a vast tub of waste, leaving at the bottom of the pipe a layer of yellow powder. This is uranium oxide, U_3O_8, or yellowcake, as it is known in the trade.

Yellowcake is the form in which uranium is usually transported and traded. Scraped off the pipe and put into containers, it is a soft powder, with a canary colour of extraordinary brilliance.

* * *

Ninety per cent of the nuclear power plants in the Western world now use enriched uranium, so most of the yellowcake goes to an enrichment plant before being fashioned into fuel rods. There, it is converted into a gas, uranium hexafluoride. The process of enrichment – that misleading word (see chapter 2) – has been described above: the two methods of enrichment, and the relationship between low enrichment for power reactors and high enrichment for bombs. An enrichment plant is huge, often a half-mile long; the process itself, whether diffusion or centrifuge, goes on invisibly, in machines contained in metal. Constant monitoring is an important part of the work: for one thing, uranium hex is extremely corrosive.

ERDA is still the world's biggest enricher, with two plants and a third under construction; it enriches uranium for American and overseas customers. If an electrical utility has a nuclear power plant and wants uranium enriched, as most do, then there are not many other places it can go to. Urenco takes orders for the British plant at Capenhurst in Cheshire, next to the gaseous diffusion plant that enriches uranium for Britain's nuclear warheads, and the Dutch plant at Almelo, and the one under construction in Germany. Eurodif, the Franco-Spanish-Italian-Belgian organization, has that enrichment plant in Southern France and another under construction and is taking orders for 1983. The Soviet organization Sovtech offers enrichment, and tells customers it will charge 10 per cent less than the market price in the West. And that is all there is.

Enriched or not, the uranium still has to be turned into fuel. If it is natural uranium fuel, the yellowcake itself is subjected to a chemical process that turns it into uranium dioxide, or U,O_2, which is a black powder. If it is enriched uranium fuel, then the uranium hex that emerges at the end of the enrichment process with its higher U-235 content is turned into the powder by cooling. The powder is compressed into pellets and roasted until it solidifies, and then the pellets are threaded together to make up fuel rods. Most countries with nuclear power plants import either yellowcake or enriched uranium and make up their own fuel rods, but there are factories in Belgium, Italy and Sweden which perform this service for customers.

Shipped to a nuclear power plant, the fuel rods are usually stored vertically above the reactor; a total fuel load consists of several thousand rods. Most reactors are two-storey buildings in which each storey is 40 feet high. On the top storey there is a machine about as tall as the reactor itself which holds the rods, and lowers them into the holes provided. Then the neutrons pouring out from the fission of U-235 atoms penetrate the neighbouring rods, and join in the chain reaction, contributing to the heat which turns water into steam which drives an electric generator.

The fuel rods are left in the reactor for three or four years, the period varying from one reactor to another. During this time the alchemy of nuclear fission changes them. Some 3 per cent of the uranium is turned into other elements, such as plutonium, ruthenium, americium or curium. Some of these elements are highly radioactive, so the fuel rods are radioactive as well as hot, and must be handled by remote control.

It is what happens to the uranium and its by-products after it leaves the reactor which is most at issue today. In the early years of nuclear power, there was no question about what would be done with them once a network of nuclear power plants was built and operating. They would be put into storage tanks for several years while the temperature fell by a few hundred degrees, and they became less radioactive. Then they would be reprocessed, that is, broken down into their constituent elements. The 97 per cent that is uranium would be extracted, and some of it used again as fuel. The plutonium would also be extracted for use as fuel. Some of it would be fed into thermal, or ordinary, reactors and some would be saved until the breeder reactors were built, so that it could be used as fuel in these and produce more plutonium. The small amount of remaining radioactive matter would be put into lead or concrete casks and buried underground or else at sea. This recycling of valuable material would undoubtedly earn the approval of conservationists.

This is a part of what is today called the closed fuel cycle, and most conservationists, who are in favour of recycling everything else from paper to body waste, are very much against it. So are a number of other people, including part of the US Government. This is because it produces plutonium in its pure form. If the used fuel is not reprocessed, then the plutonium is locked up in a hot, radioactive chunk of metal, and cannot be used. If it is extracted, it can be made into nuclear bombs.

Since it was said at one time that this plutonium could *not* be used for making nuclear bombs, this requires some explanation. The plutonium normally used in making nuclear weapons is the isotope 239. This is the first isotope of plutonium that is created in the process of uranium fission. When plutonium is created specifically for use in bombs, the reactor is shut down and the fuel rods are removed after about four months, and the plutonium, which is mostly plutonium 239, is extracted in a military reprocessing plant. This follows the guidelines laid down in 1942 by Ernest

Lawrence and carried out at Hanford, Washington. If the fuel rod is left in longer and the bombardment of neutrons continues, then more of the plutonium absorbes more neutrons and becomes plutonium 240, or 242. These are not fissile in the accepted sense, that is, they do not fission on the impact of a slow neutron, so they are not nuclear explosive material. However, like atoms of all plutonium isotopes, they fission spontaneously, albeit at a slow rate, and so produce neutrons that can create premature fission in the p-239 nuclei. In the normal operation of a reactor for power production, the fuel rods are left in for three or four years, so most of the plutonium consists of the higher isotopes. IAEA inspectors are alert for any premature removal of fuel rods.

In the early days of nuclear power, it was thought that if the plutonium assembled to make a bomb contained more than a very small amount of pu-240 or 242, this premature fission of pu-239 nuclei would ensure that the material would fizzle, rather than taking place all at one instant in an explosion. This was the basis of the idea of 'denaturing' plutonium that was talked about in the early years of nuclear energy: leaving it in the reactor so that most of the p-239 is turned into higher isotopes. In the years since then, it has become clear that plutonium that is mostly 240 and 242 (there is always a mixture of isotopes) *can* be made into a nuclear bomb. It is more difficult than it would be with pu-239. The implosion device that brings the sphere of plutonium together sharply must work more rapidly by several orders of magnitude, bringing it together in a few millionths of a second, and the resulting explosion will not be as powerful as it would be if pu-239 were used, nor as predictable in its power. Nevertheless, it could be done.

This was stated in 1972 by Dr Carson Clark, the Director of the Theoretical Division of the Los Alamos Laboratory, at a conference of scientists and strategic analysts:

> I would like to warn people concerned with such problems that the old notion that reactor-grade plutonium is incapable of producing nuclear explosions has been dangerously exaggerated. This observation is of course, of no practical interest to the United States and the Soviet Union, who have adequate supplies of weapons-grade uranium. To someone having no nuclear weapons at all, however, the prospect of obtaining weapons – even of an inferior or primitive type – could present a different aspect.

He did not say it, but the evidence on which his finding was based was not merely theoretical. Some time earlier, the AEC built a bomb with plutonium from a power reactor, having a high proportion of p-240, 241 and 242, and exploded it, just to see whether it could be done. This was one of a series of underground bomb tests in Nevada in the mid-1960s, but the information was not declassified until 1977.

In the course of reprocessing, the fuel rod is chopped up and immersed in nitric acid, and subjected to other chemical treatment which separates out the elements. This must all be done by remote control, since the material is highly radioactive. At the end of the chemical treatment, several streams of liquid emerge. From one of these streams, uranium is extracted, and from another plutonium.

About 30 per cent of the uranium can be recycled. The plutonium can be used in thermal reactors. However, its use for any purpose is now so controversial that most countries have barred the use of plutonium in thermal reactors where uranium will do just as well. It is available as fuel for breeders.

A 1,000-MW LWR produces 230 kilogrammes of plutonium each year. The anxiety about using plutonium as fuel is at the prospect of thousands of tons of plutonium being transported from place to place and stored at many sites. The theft and misuse of a few pounds of it could be disastrous.

When it leaves the reprocessing plant, it is a liquid solution, plutonium nitrate. This is usually converted at the same site into a plutonium oxide powder, which is easier to store. A plutonium bomb is normally and most effectively made with plutonium metal, but one could also be made with plutonium oxide. The explosion, precisely timed, would compress the powder into a critical mass.

* * *

There is much argument about whether we have solved the problem of nuclear waste. But not everyone is agreed about what problem to solve. Disagreement about the closed or open fuel cycle is also disagreement about the object of nuclear waste disposal.

There are three tasks that we may set out to perform. We can go for the once-through or open fuel cycle, and find a way to dispose of all the waste material from nuclear reactors permanently, instead of re-using any of the material: then the task is to ensure that the biosphere will be protected from its harmful potential for many thousands of years. We can reprocess the fuel, save the plutonium and usable uranium for the breeder reactor, and dispose of the residue. This is what is being done with most used fuel in Europe, though not in America. It is less of a task, though not very much less; the volume of waste is only about one-thirtieth as much, but it will contain all the most highly radioactive materials. Or we can defer any decision and store the used fuel for a few decades, in such a way that neither the plutonium nor the radioactivity can cause any harm through theft or accident, and we still have the option of reprocessing it at some future date.

Whatever we do with it, the problem is piling up: there was an estimated 20,000 tons of used fuel in the world at the end of 1980.

The disposal of waste is one of those several aspects of nuclear power that were pushed aside for many years. Carroll Wilson, in the retrospective look at his AEC career quoted at the beginning of this chapter, wrote: 'Chemists and chemical engineers were not interested in dealing with waste. It was not glamorous; there were no careers; it was messy; nobody got brownie points for caring about nuclear waste. The AEC neglected the problem.' And in Britain, the Royal Commission that studied nuclear power criticized the AEA for paying 'inadequate attention' to this problem.

It was not neglected entirely, but the attention paid to it both within the nuclear energy field and by the general public was very much less than it is today. The Atomic Energy Commission carried out a study with the Tennessee Valley Authority between 1953 and 1956 of the effects of water from the Oak Ridge enrichment plant flowing into the White Oak Lake. The study, released in 1956, found that radioactivity in the water contaminated fish in the lake, stunted their growth, and reduced their life span substantially. This was reported in the *New York Times* on page 42 and elsewhere not at all. Today such an alarming discovery would be trumpeted so loudly that no newspaper reader or TV watcher could miss it.

When a used fuel rod leaves a reactor, it is dumped into a steel-lined tank of water that is next to the reactor, and it remains there for at least a year, glowing, because of a curious effect of the radioactivity on the water, with an eerie pale blue light. The water cools it, and the water and the steel sides of the tank absorb the radiation.

If, after a few years, it is decided to reprocess it, then it is shipped to the reprocessing plant in steel casks each weighing some 30 tons. After the uranium and plutonium have been extracted, the remaining liquids consist of fission products most of which have an intense but relatively short-lived radioactivity. This must be disposed of permanently.

In America and Canada fuel is not reprocessed. It is left in the steel tanks, which are called storage ponds. Currently, there are moves to build storage ponds away from the nuclear reactors, so that used fuel from several reactors can be stored in one place. This is still a temporary measure, until a permanent course is decided upon, but the temporary period seems to be stretching from years to decades. Since storage ponds were intended to be only places for fuel to cool off before going on to the next stage of the fuel cycle, some other form of storage may seem desirable. Canada's nationally-owned atomic energy company has built some experimental steel and concrete silos in the Canadian Arctic, each of which holds 50 tons of spent fuel, and these show no sign of leakage.

Long-term storage, whether of all the spent fuel or the residue left after reprocessing, means really long term. In considering nuclear power, political leaders and officials often have to think in a different time frame to the one they are used to, well beyond their term of office. Because of long lead times, they must estimate economic costs and benefits over twenty-five to thirty years, and proliferation dangers over several decades. They must think about waste disposal, not in historic time periods, but in geological ones, in tens and hundreds of thousands of years.

The most radioactive elements in nuclear waste have relatively short half-lives. Because their radiation is intense, it diminishes quickly. It is reduced by more than 90 per cent after three years. After five hundred years, they contain no more radioactivity than a deposit of uranium-bearing pitchblende, and no one considers moving house to get away from the radiation of one of these.

However, the waste material can cause harm long after this. The long-lived radioactive elements give off alpha radiation, which has a range of only an inch or two, but is dangerous if ingested into the body. Some actually seek out the human organism because they are similar chemically to other elements which play a part in the life cycle: strontium 90, for instance, behaves like calcium and joins with calcium in settling in human bones. Strontium 90 and cesium, which has similar properties, have half-lives of less than a hundred years. But plutonium 239 has a half-life of nearly 25,000 years, and neptunium 237, which is created by the decay of other elements, one of two million years. If a minute quantity were to escape from its storage or burial place and find its way into the water or air, it could cause harm to humans. If some particles were to get into the soil, they could get into plants, either those eaten by humans or those eaten by animals that humans in turn eat, and so find their way into the human body. However, scientists who are handling waste management say that after some thousands of years, substances released would be no more toxic than substances released into the atmosphere by other industrial processes, such as coal mining (radium is found in coal ash), and some fertilizers.

One option that is still being considered for long-term disposal is putting the stuff in lead caskets on the ocean bed. By the time the caskets leak, so the theory goes, the remaining radioactive materials will be dissipated by the waters. However, more studies of ocean currents must be carried out before anyone comes down firmly for this. For a brief time, there was talk of shooting radioactive waste out into space, but a cursory consideration of the cost and the safety factor ruled that out.

It is necessary to keep it isolated, and to keep it dry. A fluid is always mobile, and if part of the radioactive waste contaminates a liquid, that liquid will move at some time up to the surface of land or sea. Not quickly, perhaps, but there is plenty of time. For this reason, the currently favoured plan is to solidify it in either a glass or ceramic block. This does nothing to reduce the heat or radioactivity, but it keeps it beyond the reach of fluid or moisture. The process was worked out at Harwell, with total success so far as the Harwell scientists are concerned. France has an operating vitrification plant at Marcoule, the site of the first French plutonium-producing reactors, and experimental plants are being built at Windscale and in Germany.

Eventually, the vitrified waste will be buried, but it must remain in containers on the surface for fifty years. If it were buried before that time, the heat might splinter the rocks.

Many geologists say there is no difficulty about isolating material from the biosphere for hundreds of thousands of years. Underground salt deposits, they say, are stable and waterproof, and vitrified waste buried in them will be dry and immobile for a million years, whatever changes take place on the surface. Some geologists say deep trenches under the ocean are equally feasible as the final dwelling places. These views are not undisputed, even within the scientific community.

They are certainly not undisputed outside it. There is usually local opposition to exploring for a burial site, with strong support from anti-nuclear campaigners. Nuclear power authorities complain, understandably, that these accuse them of creating a radioactive waste problem they cannot solve, while at the same time doing everything they can to prevent them from solving it. In the United States, the issue has set off a whole series of arguments over jurisdiction between county, state and federal governments. Several states have passed laws forbidding the burial of radioactive waste on their territory, and some ports have passed ordinances forbidding it access to their harbours.

As Carroll Wilson pointed out, all the parts of the nuclear fuel cycle are interdependent. A reduction in the output of uranium mines could reduce the amount of nuclear power available. Failure to find an acceptable way to dispose of radioactive waste could mean a back-up all the way through the cycle, limiting the amount of nuclear power that can be produced and so the demand for uranium. The waste issue may have a far-reaching effect on nuclear power in Germany, because of a 1977 court ruling. A law passed a little while earlier makes it obligatory for any company to demonstrate that it can dispose of its waste products before it undertakes an industrial process. The 1977 ruling said this applies also to nuclear power. Since there is as yet no long-term storage site in Germany, German waste is sent for storage or reprocessing to the French company Cogema. But Cogema's storage space is not unlimited. One German scientist said: 'This court decision is a requirement for a perfect solution to a situation, which is a weakness of us Germans. It could kill nuclear power in Germany.'

Chapter 11

NUCLEAR CONFLICT

When Jimmy Carter was elected, he followed through his campaign speeches on nuclear power by appointing four of the authors of the Ford/Mitre report to posts in his administration. Most importantly, Joseph Nye, the Harvard political science professor, went into the State Department to shape nuclear exports policy. Harold Brown became Secretary of Defence. Spurgeon Keeney, the Chairman of the group that produced the report, joined the Arms Control and Disarmament Agency. George Rathjens, Keeney's old touch football partner, went into the State Department later to deal with proliferation problems.

Carter announced his policy on nuclear power in April 1977, only two months after his inauguration. Not surprisingly, it followed the lines of the Ford/Mitre and Pan Heuristics reports, and, for that matter, of President Ford's speech the previous October. He said in his policy statement that he believed the risk of proliferation 'would be vastly increased by the further spread of sensitive technologies which entail direct access to plutonium, highly enriched uranium, or other weapons-usable material'. He said that commercial reprocessing in the United States would be deferred indefinitely, which simply confirmed President Ford's suspension of it.

The policy on nuclear exports was confirmed and made more rigid by an act of Congress, for the anti-plutonium campaigners had the ear of a number of members of Congress. The Nuclear Non-Proliferation Act passed in 1978 decrees that no spent fuel of US origin should be reprocessed without US permission, and suggests that this permission should not be given lightly. It says that US nuclear materials of any kind may be sent only to countries which accept full-scope safeguards – that is, international inspection of all nuclear facilities. It obliges the Administration to re-negotiate the 1958 US-Euratom agreement on the supply of nuclear fuel in order to insert in the treaty the requirement of no reprocessing without prior permission, and allowed two years for the completion of this re-negotiation.

The act, one of the longest and most complicated ever passed by Congress, is one of those that embodies the American constitutional doctrine of checks and balances. The President may waive its requirements if he decides that national security or non-proliferation interests require it; but Congress can overrule his waiver if both houses agree to do so by a majority vote within sixty days.

Canada, seeking to tighten its export controls after the shock of the Indian explosion, also set out to re-negotiate its uranium supply contracts with Euratom and Japan, to inject clauses stipulating no reprocessing of Canadian uranium without prior consent. Both Japan and Euratom refused to re-negotiate their contracts to add new restrictions, so Canada eventually halted shipment of uranium to both, though it resumed it when negotiations began again. Both protested at this breach of contract.

Europeans and Japanese reacted angrily to the US and Canadian policies, and others did also. As they saw it, the United States was trying to use its position as the principal supplier of enriched uranium to dictate to the rest of the world the way it should run its nuclear power programmes. Even many of those who agreed with the basic anti-plutonium thrust of the policy objected to the style. The accusations that were made by Third World countries against the London Suppliers' Group were now being made by some members of the group against others.

The American and Canadian policies touched a sensitive nerve. The 1973 oil crisis and subsequent price rises were a shock for the United States, but a bigger one for Europe and Japan, which have far fewer natural resources. Japan, a densely populated and heavily industrialized island nation, is even more vulnerable to an interruption of energy supplies and other raw materials than Western Europe. Japan went to war in 1941 because its supplies of raw materials were threatened. Most European countries depend on the Middle East totally for their oil. (Hence the anti-Israeli tilt of EEC statements on the Middle East in recent years.) The European Commission urged countries to reduce this dependence, partly by reviving coal mines wherever possible and developing new sources of energy, and partly by expanding their nuclear power programmes. In planning ahead for decades, this seemed to point to a fast breeder programme.

But the US rules would mean either abandoning any plans for such a programme, or else continuing it on American sufferance, with each batch of used reactor fuel going for reprocessing and the extraction of plutonium for fuel only with American permission. Western Europeans did not want to exchange an uncomfortable dependence on the changing policies and goodwill of the Arab countries for what was coming to seem like an uncomfortable dependence on the United States.

Joseph Nye appeared to show a recognition of this feeling of insecurity in the title of an address he gave to an international symposium of the Uranium Institute in London in 1978. His talk was to be called simply, 'Non-Proliferation: the American Position'. But at the last moment, he changed the title to 'Balancing Non-Proliferation and Energy Security'. However, he showed an insensitivity to these same anxieties in the course of the talk, when he said the Nuclear Non-Proliferation Act is not as strict as some people seem to think because the President can overrule a refusal to issue an export license, as if this provided reassurance. A German in the audience said afterwards, 'Does he really think we're going to let our industrial life blood for the next forty years depend on the whim of some future American president?'

Euratom countries were confident when they turned to light water reactors and hence dependence on the United States to enrich uranium for them. Now this confidence had vanished. Bertrand Goldschmidt, the French physicist who has worked on nuclear energy since the wartime bomb project and was for a long time the CEA's Director of International Relations, illustrated this new discomfort in a passage in *Les Rivalités Atomiques*, a book he wrote shortly after he retired from his official post: 'Who knows whether tomorrow, a nation, even though it has respected the clause on peaceful uses in a bilateral agreement, will not find itself deprived of the fuel it needs for its reactors for purely political reasons, perhaps even reasons not connected with atomic energy?'

The US policy may prove to be counter-productive. It may spur other countries on to attain nuclear self-sufficiency, by developing their own enrichment capability and possibly going through the reprocessing-breeder cycle. Justice William Fox, the Australian who served for a while as ambassador-at-large to deal with uranium exports, warned Nye of this at lunch at the Australian Embassy in Washington. Fox was also worried about plutonium, and the use to which his country's uranium might be put, but he thought the American policy heavy-handed.

Talking to Europeans involved in nuclear power during the period after the pronouncement of this policy, two images emerge. One is that of Jimmy Carter as a kind of barefoot Woodrow Wilson, a blinkered moralist, well-intentioned but so blunderingly unworldly as to be dangerous, trying hamfistedly to impose home-grown American solutions on the rest of the world for the world's own good. The other is also a familiar image of America encountered abroad, and a more malign one: of greed and self-interest masquerading as global high-mindedness.

For many Europeans see commercial motives behind American policy. These usually point out that while the United States is the world's principal supplier of nuclear reactors and of enrichment services, its attempts to construct a commercially viable reprocessing plant have not so far been successful. This is one service that Britain and France can provide that the United States cannot, and one that other countries also might create. Officials and people in the nuclear power industry in several European countries express the view privately that the American aim is to prevent other countries from developing services that American industry cannot sell. A Belgian official said darkly, 'Just you wait. A few years from now, the Americans will have caught up with our reprocessing technology, and then they'll come out in favour of commercial reprocessing.'

When this view, albeit expressed more delicately, emerged during some diplomatic negotiations, an American tried sarcasm in response. 'Have you noticed,' he asked, 'how the American nuclear power industry is prospering under the Carter policy? Have you noticed how enthusiastically it supports these policies?'

For the American nuclear power industry was not prospering, and it did not support these policies; it opposed them, often heatedly. One leading figure described the Carter Administration at an industry meeting as 'a bunch of eco-freaks'. Assailed from many sides now and accused of perpetrating great evils, many people in the nuclear power industry are developing an understandable paranoia, and regard anyone who tries to restrict any nuclear power activity as the enemy. They tend to identify opponents of the reprocessing/breeder cycle with all-out campaigners against nuclear power, or at any rate, to view them much as conservatives view Socialists in relation to Communists: advocates of a milder form of the evil represented in an extreme form by the other, bad both in itself and as a probable pathway to something worse.

Views on nuclear power can make some strange bedfellows. A conference on the transfer of nuclear technology organized jointly by the IAEA and the OECD was held at Persepolis, Iran in March 1977. The Third World was represented strongly and vociferously, as was to be expected. The developing countries issued a statement deploring the new restrictions that were being placed by industrial nations on the transfer of nuclear materials. It was written by the President of the American Nuclear Society, the nuclear power industry's association.

* * *

Britain alone of all the European countries does not rely on America for enrichment. Its first generation of nuclear power plants, the Magnox reactors, use natural uranium which does not require any enrichment. The AGRs use 2 per cent enriched uranium, but by the time the first of these was built, the Urenco enrichment plant at Capenhurst was operating.

The experimental fast breeder reactor at Dounreay in the North of Scotland is the longest-operating breeder reactor in the world, and now a larger one has been built next to it, and the AEA look ahead to a breeder future. The Flowers Report in September 1976 came as something of a bombshell. This was the sixth report of the Royal Commission on Environmental Pollution, on Nuclear Power and the Environment. The Commission was headed by Sir Brian (now Lord) Flowers, the Rector of the Imperial College of Science and Technology, a physicist who came out of university into a minor role on the wartime atom bomb project. Sir Brian gave a wide interpretation to his brief, and took it to include just about every aspect of nuclear power – after all, a nuclear explosion, whatever else it did, would certainly damage the environment.

Sir Brian Flowers was, and is today, a Deputy Governor of the Atomic Energy Authority, but his commission's report contradicted the received wisdom of the nuclear power establishment. It foresaw many dangers in a nuclear power cycle that relies on plutonium as fuel, such as the danger that terrorists might seize some plutonium: 'Plutonium appears to offer unique and terrifying potential for threat and blackmail against society.' The report came down in its conclusion against bringing in the breeder reactor soon, though not conclusively, because it recognized uncertainties: 'Our anxiety about the hazards of an economy based on plutonium leads us to the view that fast reactors should be introduced only if they are demonstrably essential. . . . Our alternative strategy suggests that fast reactors might be avoided at least over the fifty-year period covered.'

The report also criticized the AEA's attitude to nuclear waste: 'We think that quite inadequate attention has been given to this matter.' It expressed confidence that a safe, long-term method of disposing of waste could be found, but said that it has not been demonstrated yet, and went on: 'We are agreed that it would be irresponsible and morally wrong to commit future generations to the consequences of fission power on a massive scale unless it has been demonstrated beyond reasonable doubt that at least one method exists for the safe isolation of these wastes for the indefinite future.'

Not many people in Britain have concerned themselves with questions about plutonium and breeder reactors, either in the Government or out of it, so most pronouncements and most debate takes place either within a small and specialized group, or else publicly in a formal structure. Following the Flowers Report, there was a big debate at which all the arguments about reprocessing and plutonium were raised. This was the Parker Inquiry in 1977, an inquiry under the Town Planning Act into a proposal by British Nuclear Fuels Ltd., the nationally-owned nuclear materials organization, to build a big new reprocessing plant at Windscale, on the Cumbria coast, alongside the one already in existence.

Fuel rods from the Magnox reactors have always been reprocessed at Windscale. The Magnox fuel rods must be reprocessed; they cannot be stored in water for long because the cladding will corrode and the dangerously radioactive materials inside will spill out. The proposed new plant, which would be very much bigger, would reprocess the thermal oxide fuels from the AGRs, so it is sometimes called THORP – standing for Thermal Oxide Reprocessing Plant. Half of the plant would be given over to reprocessing enriched uranium fuel from LWRS overseas, and half of this would come from Japan.

The Inquiry lasted 3½ months, and was the longest such inquiry ever held. Massive testimony was given on both sides. All aspects of the case for and against the reprocessing/breeder cycle were restated: that it can make nuclear weapons more accessible, that it does or does not make disposal easier, that it will or will not be economic. Albert Wohlstetter came from America to testify for the opposition to THORP. A few witnesses argued the case against nuclear power of any kind. BNFL admitted that there had been a number of leakages of low-level radioactive materials from the storage tanks. An Irish academic argued that since low-level waste was being poured into the Irish Sea, and since any radioactive liquid or gas released accidentally was likely to go into the Irish Sea or over it, Ireland was at risk. Others argued that nobody was at risk; a trade unionist representing Windscale workers said that more people would be killed and more pollution caused by cars imported from Japan than by used nuclear fuel imported from Japan.

Joseph Nye brought the US Government into the situation by sending what he called a letter of clarification to the Inquiry. This said it would be incorrect to interpret certain evidence as indicating American endorsement of the THORP project. This could only be seen in Britain as opposition to it, and conspicuous interference. In fact the American Government was in a *de facto* alliance with the opposition to the new facility, just as, in the early 1960s, when the Campaign for Nuclear Disarmament was in full swing in Britain, the State Department and the Pentagon, which did not want to see a continuation of Britain's independent nuclear deterrent, were in a *de facto* alliance with the CND marchers.

A significant argument was advanced by BNFL in the course of the Inquiry. This was that it was safer from a proliferation point of view that countries should send their used fuel for reprocessing to Britain, a country that already has nuclear weapons and therefore has no need to divert plutonium, rather than that they should build reprocessing plants themselves. This argument is underlined by a clause in BNFL's contract with Japan that said that the plutonium extracted during the reprocessing will only be sent back to Japan when Japan can demonstrate that it is required for fuel.

The Inquiry gave the go-ahead to the Windscale expansion, and the new plant is due to be completed some time around the end of the 1980s. Most Britons saw this purely in domestic terms. They welcomed the boost this would give to Britain's trade figures, or else worried about a part of Britain becoming a 'nuclear dustbin', in the popular phrase coined by a headline writer. But the *New York Times* commented on Justice Parker's decision: 'Britain is about to undermine President Carter's campaign to curb the spread of nuclear technologies that can be used to make bombs.'

The Windscale expansion plan was in fact predicated on the collapse or reversal of the US policy. This is an aspect that has never been brought out. When the reprocessing plant is built, only half the intake of used fuel will be from British reactors. The other half will be from foreign reactors, and the largest part of this will come from Japan. Japan is financing the construction on a cost-plus basis. Most of the foreign nuclear power plants that would send their used fuel to Windscale – in Spain and Switzerland, for instance – are PWRS, and use uranium enriched in the United States. All Japan's do except one, the British-built Magnox, which uses natural uranium. This means that they will have to get American permission to send the used fuel to Windscale for reprocessing, and also Canadian permission where the uranium was bought in Canada, which much of it was.

If the US Government refuses this permission, then half the plant will be standing idle, and the huge sum spent on its construction will have been largely wasted. Japan will not be able to send its used fuel there. But there is no provision for Japan to get its money back from BNFL in such circumstances. Evidently, Japan does not think this will happen.

* * *

The arguments about reprocessing and breeding come back in the end to the availability of uranium. If uranium is plentiful and readily accessible then there is no hurry about building breeder reactors and recycling fuel. It will not be necessary and it will not be economic.

In May 1977 David Owen, then British Foreign Secretary, gave a speech on non-proliferation, and addressing himself to the breeder/plutonium issue, he gave primacy to the uranium question:

> Are uranium reserves in non-Communist countries sufficient to meet the demands of reactors planned for construction by the mid-1990s? If so, for how much longer? The Ford/Mitre Study thought that at projected rates of demand, there would be no difficulty about fulfilling requirements for uranium up to the turn of the century. Other authorities take a different view. Either way, what happens after the turn of the century?
>
> Will uranium ore follow the same general history as other minerals? That is, with foreseeable advances in technology and changed economic conditions, will it be possible to exploit ores and reserves of progressively lower grades and lesser accessibility? Once upon a time, no one would have touched copper below a yield of 20 per cent. Today half-per cent copper is economic in some parts of the world. In the latter half of the nineteenth century, no one would have touched the lower grade ore which we now mine in the United Kingdom. How much difference will higher prices (and what prices?) make to the availability of uranium and the economics of nuclear power generation? The answers to these questions will greatly affect the policy conclusions we draw.

There is room for doubt about the availability of any mineral, and more in the case of uranium than most. It is only in recent times that the extent and scope of Man's activities has brought the finite nature of the Earth's resources into view. When estimates are made of mineral resources, these tend to under-estimation. This has certainly been true of uranium ever since the 1944 report to President Roosevelt that there was only enough uranium in North America to build the wartime atom bombs, and no more. The original impetus for the atoms-for-peace programme was the belief that, if the major powers gave over a quantity of uranium for peaceful uses, they would not have enought left over to build large numbers of nuclear weapons. In 1966, the AEC estimated US reserves of uranium at 195,000 tons. Some 130,000 tons was extracted in the next ten years, and then, in 1976, the reserves were estimated at 430,000 tons.

Of course, it is more sensible to base present policy on the best possible present-day estimates of uranium resources rather than to extrapolate from past mistakes and assume that any published figure today understates the position just as past ones did. But the published figures even of geologists provide no certainty, and still leave room for doubts and judgments.

The Ford/Mitre report concluded that there is a plentiful supply of uranium, enough to fuel nuclear power plants beyond the end of this century without fast breeders, and this was an important pillar of its case against the breeder. The S.M. Stoller Corporation, a research organization dealing in resources, conducted a study for the US Department of Energy, and came to a contrary conclusion. It says that given median nuclear power growth, there is only a fifty per cent chance that domestic uranium reserves will be able to meet US needs during the decade 1990–2000, and a smaller probability that it will meet them in the following decade. The Stoller Corporation considered the workings of the uranium mining industry in a way that the Ford/Mitre Group had not, and cites limiting factors: the increasingly low grade of ore; the fact that most uranium mines are now in the hands of big companies which have competing demands for money available for investment; and the difficulty of finding more uranium miners. Department of Energy officials, arguing against this view at meetings, retort that the amount of investment required for uranium mining is small change compared to the amount that would have to be found for the expansion of nuclear power that would use this uranium, and as for miners, if more men are not going to be found to blast out uranium, then many more will have to be found to dig out coal.

The most broadly-based single survey of uranium supplies is the one published occasionally by the IAEA and the OECD, called *Uranium Resources, Production and Demand*. It covers the entire non-Communist world. The one published early in 1980 takes issue with the Ford/Mitre Group, though not explicitly. It says that demand will overtake supply in the 1990s unless breeder reactors are introduced, even accepting a wide range of uncertainty about the amount of nuclear power that will then be in use. It estimates that deposits of uranium that can be mined at a cost of up to $80 a kilogramme (slightly less than $40 a pound) total 1.85 million tonnes. This is 200,000 tonnes more than the figure given in the previous report, published two and a half years earlier. It estimates that there are another three million tonnes that can be mined at a cost of $130 a kilogramme.

Some other studies support this conclusion, others contradict it. The British Department of Energy published a paper predicting a heavy demand for uranium and steeply rising prices. The Science

Policy Research Unit at the University of Sussex published one forecasting new discoveries, a glut of uranium and falling prices. Two American geologists, Kenneth S. Deffeyes and Ian D. MacGregor, produced a book-long report for the Department of Energy that reached quite different conclusions to the S.M. Stoller Corporation's report to the department. Theirs, which was published in a shortened version in the *Scientific American*, draws on several years of research into the patterns of discovery of uranium and other minerals, and suggests that enough uranium will be available for the next hundred years of nuclear power without the breeder.

The Uranium Institute, an international organization of uranium producers and users, took note of these discrepancies in a report and observed: 'US production estimates by the Department of Energy in recent years have consistently exceeded those made by independent analysts working in the private sector.' The Institute goes on to offer an explanation of this: 'The difference must in general terms reflect the gap between what could be achieved in ideal circumstances, and what is likely to be achieved in the real world of investment decisions, regulatory constraints, and a market which is being progressively opened to foreign uranium.'

Some people in the nuclear power industry opt for a simpler explanation: that the Department of Energy cooks the figures, or at the very least, is biased heavily on the side of optimism. Optimism about uranium supplies supports the contention that there is no need to hurry into breeders.

Uranium exploration doubled in the last three years of the 1970s according to IAEA figures, despite falling prices and the fact that there is already a buyers' market. Utilities and nuclear power organizations want access to supplies. Britain's Central Electricity Generating Board, the French CEA, and American, German and Italian companies are financing exploration in many parts of the world. Uranium deposits lay behind the murky story of the French Government's relationship with the infamous Emperor Bokassa of the Central African Empire. A company owned indirectly by the French Government, Amok, is mining the deposit at Cluff Lake, Saskatchewan, which has the highest grade uranium ore discovered anywhere in the world.

In the United States, the Department of Energy is carrying out a seven-year National Uranium Resource Evaluation, which is to go on until 1984. It hopes this will turn up some new kinds of deposits away from the sandstone and carnotite areas of the Southwest. Already some American companies are experimenting with the extraction of uranium from potash in Florida.

The EEC Commission is financing uranium exploration in Ireland, Sardinia, and several other parts of Europe. Uranium deposits have been located which are not yet being exploited, in Mexico, which may prove to be particularly important, and Algeria, Argentina and Brazil. The mining of substantial quantities of uranium in these countries would be of more significance than new discoveries in Canada and Australia, because it would represent a diversification of sources of supply, and would reduce the extent to which big users are in thrall to a very few countries.

Uranium is still being found in different kinds of ore. The granite deposit at the Rossing mine in Namibia is unique so far. So is the phenomenally rich ore at Cluff Lake, where the uranium is concentrated in gaps that appeared because of a fault when rocks formed in layers 2,500 million years ago.

Exploration today is a far cry from the lone fortune-hunter with a geiger counter hiking through the Rockies or bouncing along in a jeep across the great Australian desert. It may be a mission financed by a Government department or a big corporation a continent away. Two men in a light plane, with a spectrometer attached to the fuselage sensitive enough to detect even the small amount of gamma radiation coming from uranium, fly low over the ground looking for irregularities in the pattern of radiation, which they call 'anomalies'. If they find one, and it is in an accessible place, a team on the ground goes to the spot and samples the surface ore. It may be that the gamma radiation comes from some other radioactive substance. However, if it looks as if there might be uranium, they bring in a special drill and bore down 2,000 feet, and lower a gamma ray detector down the hole. If it registers, then they drill another hole a few feet away, down to the depth at which gamma rays were detected, and lower a detector again. If gamma rays are detected here also, then they use a core drill to bring up a sample of rock from that depth, and analyze it for uranium. Decisions on whether to go ahead after this are taken at some corporate headquarters.

Several countries are experimenting with the extraction of

uranium from sea water. Uranium is in all sea water, but only in a proportion of three parts to a billion on the average, so that extracting it is extremely difficult. The UK Atomic Energy Authority had a chemical study going for some years but it proved unpromising. The Japanese Government is financing the construction of a $12 million pilot plant on the Southern island of Shikoku. Officials who are pressing the scheme say this will one day produce uranium at $250 a pound, but they have offered no evidence of this, and in any case, this is about eight times the present price of uranium. However, price is not the most important consideration in these experiments. As the director of a Government-funded sea water project located near Hamburg explained, in a lecture on his work: 'With nuclear power generating capacity installed, an independent and secure uranium supply even from exotic and expensive sources is to be preferred to reasonably priced but uncertain supplies.'*

The issue of uranium must always come down to price, because this is the limitation. There is no question of the world literally running out of uranium; the amount of uranium is virtually infinite in relation to any possible demand. As we have seen, it is abundant in the Earth's crust. But it is thinly distributed, and it occurs only here and there in concentrations. The question is not whether there is uranium, but how much it costs to extract it. Estimates of reserves always have a price attached to them: X tons are estimated to exist at $40 a pound, Y tons at $80 a pound, and so on. Clearly, there comes a point at which the uranium in the ground (or water) is so costly to extract that its use is no longer a solution to the energy problem, since it is more expensive than whatever remaining supplies there are of oil and coal. A theoretical limit is the point at which the extraction of the uranium requires more energy than the uranium could produce.

However, there is a great deal of elasticity in uranium prices. Nuclear power, in contrast to coal or oil power, is capital intensive. The biggest cost by far is the cost of building a nuclear power plant. The fuel is a much smaller part of the cost. In a PWR, the uranium today is about 4 per cent of the cost of operating it while the capital cost is being amortized, and the enrichment of the uranium is

* Extraction of Uranium from Sea Water by Peter H. Koske, published in *Uranium Supply and Demand*, 1979.

another 6 per cent. A thousand megawatts of PWR electricity requires 350 tons of uranium a year; 1,000 megawatts of electricity from a coal-fired power plant requires 2½ million tons of coal, which means that transport is a major part of the cost. The increase in uranium prices by a factor of eight in the early 1970s had little effect on the cost of nuclear electricity, while the quadrupling of oil prices transformed the economics of power and transport. We could pay a lot more for uranium without any great effect on our electricity bills.

A breeder reactor is more expensive to build than a thermal reactor. The CEA says that the Super-Phénix, the 730-MW fast breeder that is being constructed near Lille and is due to start operating in 1983, will produce electricity at twice the cost of the Framatome-built PWRS. The commercial justification for the breeder is that it will save fuel. The AEA says the breeder will become economic in Britain when the price of uranium goes up to $80 a pound, and the AEA expect it to do so by the time that the first few breeders are operating, in the last years of the century. Sceptics say that no full-scale breeder has been built yet in Britain or anywhere else, and forecasts of what a nuclear power plant will cost have usually been below the actual cost. The UKAEA says it has now had enough experience with prototypes to be able to scale upwards and calculate the costs accurately. The usual way of assessing the economics of a breeder is to give the price of uranium at which it will be cheaper to operate the breeder than to buy uranium and operate a once-through cycle. A recent study of the fast breeder by the Hanford Engineering Laboratory concluded that this price will be somewhere between $74 and $258 a pound. The wide range indicates the range of uncertainty in cost of the fast breeder programme. Others have given higher figures still for the break-even point of the breeder, for instance Brian Chow in a report prepared for the Arms Control and Disarmament Agency (though this is regarded with some suspicion by breeder partisans, because it comes from Pan Heuristics, Professor Wohlstetter's think-tank, and was commissioned by ACDA, which is institutionally more concerned with proliferation dangers than with energy availability).

The plutonium that will be fed into breeders will come from a reprocessing plant, and the cost of reprocessing used fuel has increased gigantically. At one time, the US price was $30 a pound. A senior figure in the American nuclear establishment told an international conference in 1977 that he thought the price would come down until the cost was one dollar a pound. Currently, the French company Cogema charges its new customers $227 a pound.

However, fast breeders may also help solve the waste disposal problem by reducing the volume of waste, even though the amount of radioactivity will be very nearly the same.

Walter Marshall, the portly, blunt-spoken Yorkshireman who became the Director of the UK Atomic Energy Authority at the beginning of 1981, has put forward another argument for fast breeders that is taken seriously. He challenges the Ford/Mitre-US Government position on its own ground, and says breeders will actually reduce proliferation risks and the danger that plutonium will be misused. He points out that the radioactivity in used nuclear fuel is intense but therefore short-lived, and decreases rapidly. After a year, he says, 99½ per cent of the gamma radiation has dissipated. This means that as time goes by, it becomes easier to handle, though it is still not without its dangers, and a government or a well-equipped terrorist organization would be more able to extract the plutonium for its own purposes. In the breeder cycle, however, plutonium is extracted from the used reactor fuel and re-used quite soon, and it can remain mixed with highly radioactive fission products even while it is fabricated into breeder fuel rods along with uranium. He concludes that the safest thing to do with plutonium is to burn it in breeder reactors.

Carroll Wilson reminded us that all stages of the nuclear fuel cycle are interdependent. This is reflected in the interdependence of all the questions concerning the breeder. *If* plenty of uranium can be found and mined at the right price, then the breeder will be unnecessary. *If* reprocessing becomes still more expensive, it may be uneconomic, unless uranium becomes more scarce and more expensive still. *If* radioactive waste can be disposed of permanently and safely, so that there is no danger of the plutonium being misused, then the breeder is unnecessary *if* there turns out to be more uranium available. *If* the demand for nuclear power plants continues to fall off, then there may be no necessity for breeders even if there is less uranium than the most optimistic estimates say. And so on.

President Carter recognized other countries' anxieties about uranium supply and disposal by coupling his opposition to reprocessing with an order to speed up construction of the new enrichment plant, to provide more fuel, and also a vague proposal that the United States store other countries' used fuel. In fact, the US Administration is having difficulty finding places to store even American fuel, and meeting opposition everywhere. For a while, the United States considered using the almost deserted Pacific island of Palmyra as a storage site for Japanese fuel, but the project encountered opposition from Pacific nations and from the Japanese themselves, who found that they were expected to pay all the cost.

Of course, the appeal of the fast breeder is also strategic. Indeed, some French officials give the impression that the prospect of nuclear independence ranks so high in their thinking that the economics are of little importance so far as they are concerned. And the opposition to it is also strategic: concern that the spread of plutonium will increase the risk of proliferation.

The issue can be seen as a choice of risks. Gunter Hildebrand, of the German nuclear reactor company KWU, wrote in an article about the US restrictions: 'In the view of an importer, the risk of misuse of dependence is much greater than the danger of breaking an international safeguards agreement, which no one has yet done.'*

And a Japanese nuclear power economist, Koichi Kawakami, complained in an article that American policy seems to be 'based on the idea that the prevention of nuclear proliferation is the only and supreme objective, with all other objectives excluded or belittled.' †

American arguments against the early introduction of the breeder cycle that are put to Europeans and Japanese often boil down to: 'We don't really know how much uranium will be available, we don't know what changes technology can bring in the next twenty years. Plutonium is dangerous. With all this uncertainty, why take a chance and go for fast breeders now?'

The European/Japanese attitude often boils down to: 'We don't really know how much uranium will be available, we don't know what changes technology may bring in the next twenty years. With all this uncertainty, why take a chance. It's better to go in for fast breeders now.'

* * *

* *International Security*, May 1978

† *Bulletin of the Atomic Scientists*, June 1978

The United States cannot simply impose an anti-plutonium policy on the rest of the world, and particularly not on its allies. For one thing, it would have only a temporary effect. The American weapon is the possibility of denying enrichment capacity, but other countries can build enrichment plants and have done so. Canada is in a stronger position, as the principal supplier of a raw material that cannot be duplicated. But the amount of muscle power is not the real issue. Neither the United States nor Canada wants to play the international bully boy, even in a good cause.

America has been in retreat from its original anti-plutonium stand almost since it was promulgated. Having laid down the principle, it qualified it away in the practice.

When export licences for enriched uranium for Euratom were suspended in 1978 as required by the Nuclear Non-Proliferation Act, because Euratom refused to negotiate new terms in its contract, the Nuclear Regulatory Commission rushed through a batch of licences just before the suspension, covering the next twelve months so that the Euratom countries would have a good supply of fuel in hand.

The Administration waived the provisions of the Act to allow Japan to send several shipments of used fuel rods to Windscale and Cap La Hague for reprocessing. The reason given was that there was not enough space for the used fuel in Japanese storage ponds, or, in one case, that the local populace had been promised that no more used fuel would be stored in the locality. It gave Japan permission to reprocess US-origin used fuel at a pilot reprocessing plant at Tokai Mura, which really meant that it gave Japan permission to build and operate the plant, since the plant had no other purpose. It continued to send enriched uranium to India. In fact, it did not turn down a single request to reprocess US-origin fuel.

Anti-plutonium fundamentalists in America were upset by these concessions, Representatives Jonathan Bingham and Paul Findley wrote to the State Department in December 1978 complaining about the apparent erosion of the policy. 'It would appear,' they wrote, 'that the US guidelines are so supple as to assure reprocessing approvals. . . . At what point our accommodations begin to be construed abroad as a change of mind or a collapse of will is, in our view, a question that now must be taken seriously.'

The leaders of the countries most affected by the American

policy had an opportunity to put their objections to Carter in person just a month after he announced the policy, when they met at the economic summit meeting of seven industrialized countries in London, and they did so forcefully and lengthily. President Giscard and Chancellor Schmidt spoke out strongly. Mr Trudeau was sympathetic to the American anxieties about plutonium, and recalled for the others Canada's experience with India, in some detail. Prime Minister Callaghan supported the other Europeans, but protested at one point that too much time was being given over to this subject.

Mr Carter knew of the opposition to his policy from State Department cables, but he was surprised at its intensity. He had said in his policy speech that America would discuss with other countries a joint policy on nuclear power and proliferation, and now they all agreed readily that their governments should study together the alternative possible nuclear fuel cycles and the proliferation risk. The final communique of the conference promised a joint study.

This was the International Nuclear Fuel Cycle Evaluation, INFCE, which began with an organizational meeting in Washington the following December and went on for 2½ years. At first, these seven countries were going to carry out the study alone, but then they realized that the rest of the world would regard this as a rich men's club, like the London Suppliers' Group, and would not be likely to pay much attention to the findings. So they invited others to join in, and by the end, sixty-four nations took part in one way or another.

INFCE was useful immediately because it provided a way for everyone to back down. The Euratom Council agreed to renegotiate its nuclear supply agreement with the United States, but said it would not discuss the requirement of prior consent before reprocessing – which was the only thing of significance that there *was* to discuss – until INFCE had reported. This was obeying the letter rather than the spirit of the Non-Proliferation Act, but American officials agreed at once and fuel shipments were resumed. Canada also resumed shipments of uranium to Euratom and Japan, pending INFCE's report. When the two-year period required by the Non-Proliferation Act passed without any agreement between the United States and Euratom, the Administration granted an extension under executive powers contained in the act, and has continued to do so ever since.

Some people thought that INFCE would be this and nothing else: a cloak behind which America could withdraw from an ill-judged policy. Others thought the whole thing was a waste of time, and would benefit only the airlines and Viennese hoteliers (plenary sessions were all held in Vienna). But some Americans had high hopes for it. Gerard Smith, the chief US delegate, who has been negotiating on nuclear arms control for many years, said at the first plenary session: 'The United States would like to see reports produced that have an educational effect. We would hope that they would support US non-proliferation proposals.' He was unusually frank in saying that he wanted to see a report favouring the American position, yet his use of the word 'educational' was sincere. American participants really felt that the Europeans and Japanese were not fully aware of the dangers of plutonium spread, and would change their minds once they became aware of it. They expected to get what they wanted from INFCE, not through the normal logrolling methods of negotiations, but through educating others to their point of view.

The aim of INFCE was a technical consensus. At the planning session, it was agreed that it would be a 'technical and analytic study, and not a negotiation'. This was yet another attempt to exclude political judgement from a field from which it cannot be excluded.

Participants divided up into eight groups, each to study a different topic, and often, the division of opinion within the groups was all too familiar.

Group one dealt with fuel supply. The division there was between those whose countries planned to build fast breeders in order to meet a shortage of uranium and who therefore believed that there was likely to be a shortage of uranium, and those whose governments feel that fast breeders are dangerous and probably unnecessary for the time being because there is plenty of uranium. The Chairman of the group was Robert Nininger, an American geologist, who is the principal author of the OECD/IAEA reports on world uranium supplies. Nininger applied the US Department of Energy model to the world scene. Europeans protested that this was too optimistic, and that when he went beyond the firelight of those deposits that were already known and were being exploited into the semi-darkness of 'presumed' and 'probable' resources, the availability was much less certain than he would have them believe, and certainly not something that can be fed into long-term energy plans. In one aspect, of course, this was simply a clash between detached speculation and the more cautious assessment of those who must depend on the uranium actually being available. But it also mirrored the arguments over uranium supply that have been going on internationally and within several countries.

Working group five dealt with fast breeders. In this, the Soviet delegate presented a report on the fast breeder that contained only the most optimistic forecasts on cost and performance and ignored negative factors, so that it read more like a car salesman's prospectus than an objective assessment. As expected, the Americans were foremost among the critics, while the French went to the Russians' defence.

If the Americans expected INFCE to educate the Europeans, some Europeans expected the Americans to learn from it and broaden their view (that is, those who were not completely cynical about the whole exercise). The Europeans seem to have won. Most of the INFCE conclusions pointed to a rejection of the US/Canadian anti-plutonium policies. Insofar as US officials expected it to provide international backing for an anti-plutonium stance, they were disappointed.

In a summary of the vast INFCE report prepared for the US Senate Foreign Relations Committee, Warren Donnelly identified aptly the INFCE themes that differed from American policy:

> One is that nuclear power, instead of being an energy source of last resort, as President Carter continues to say, is as INFCE says 'an inalienable right' of all nations. . . . The other implicit INFCE theme is that plutonium is a legitimate fuel for nuclear power; that it may well be in use within the next two decades; and that all nations have a right to use it, although economics are likely to confine it to the most advanced nuclear power countries. INFCE recognizes that there are risks associated with commercial production and use of plutonium, but indicates that these risks can be kept within limits acceptable to the world community.

The report of working group three, on long-term assurances of supply, said that where there is a prior consent clause on reprocessing, the criteria for the granting or withholding of consent should be established in advance in so far as possible. It also said, in implicit criticism of the US and Canadian policies, that where an agreement on supply exists, new conditions should not be applied unilaterally, 'even if they are associated with non-proliferation objectives'.

The report bridged a lot of the disagreements about future trends by giving high and low projections, and relating these to one another: of uranium availability, electricity demand, fast breeder costs. The projections of nuclear power in the non-Communist world in the year 2000 ranged from 850 to 1,200 gigawatts, and in 2025 from 1,800 to 3,900 gigawatts. (The report said at one point: 'Nuclear capacities fifty years into the future cannot be forecast with a high degree of confidence,' a statement that, considering the record of forecasting in the field of nuclear power, points to a considerable capacity for cool understatement.) This enabled the report to say that uranium supply should be adequate without fast breeders by the end of the century if nuclear power output remains within the low range of these projections, but not otherwise.

The Chairman of the UKAEA, Sir John Hill, had seen an advance notice of the INFCE report when he introduced the Authority's annual report for 1979, and he sounded well satisfied. 'The view as of today,' he said, 'is that INFCE has to all intents and purposes confirmed the policies being pursued in the United Kingdom. We find this most reassuring.'

* * *

The Reagan Administration took a more positive view of nuclear power than its predecessor from the start. This was to be expected. Reagan's election had portended a pro-business administration, and in his election campaign, he had compared technology favourably to trees as environmental pollutants. There was no sharp turning point, no historic speech such as those by Eisenhower and Ford, but a whole lot of statements in the early months of the Administration which indicated a change of mood.

The Assistant Secretary for Energy, Mahlon Gates, told the Senate Appropriations Committee soon after he was appointed: 'I am pleased to announce that the Department has embarked on a course of action intended to reverse the deterioration of our nuclear industry and restore us to a position of leadership in the international community.' Another senior Department of Energy official, Jerry Griffiths, the Director of the Office of Light Water Reactors, promised to reverse what he called the Department's 'bias against nuclear power'.

The new Secretary for Energy, James Edwards, proclaimed himself in favour of the fast breeder from the start. 'I would like to go ahead with the Clinch River breeder reactor,' he told a Congressional committee. He went on to say that if the United States developed breeders, 'we could have enough electricity to serve this country and the free world for a thousand years to come. We are not taking advantage of it. It just doesn't make sense to me.'

More funds were allocated to the Clinch River breeder, but, surprisingly, the House of Representatives refused to give a green light to the project; in any case, the approval of the Nuclear Regulatory Commission is required before construction can start. The Energy Laboratory of the Massachussetts Institute of Technology produced a report forecasting expanding uranium production and growing stockpiles, and, therefore, a delay in the point at which reprocessing and the breeder reactor become economic, but the Administration appeared to take no cognizance of this.

On nuclear exports, the Reagan Administration has carried forward what had become the practice of the previous Administration though not the principle, as has happened in a number of other fields (arms sales, Central America). It has eased licensing procedures slightly, continued the policy of waivers and extensions on exports, and speeded up the granting of export licences.

However, it cannot abandon entirely the anti-reprocessing line. An act of Congress, the Nuclear Non-Proliferation Act, lies across its line of retreat. It can issue extensions and waivers and soften the restrictions of the Act in its implementation, but the Act is not infinitely flexible, and it must deny fuel to countries which follow certain practices. India is a case in point. For years, the Carter Administration found reasons for sending one more shipment of fuel for the Tarapur reactor, and then another, despite India's policies, while it tried to persuade the Indians to accept full-scope safeguards. Each shipment required the approval of Congress, which was achieved with increasingly narrow margins. The State Department complained that this was a constant irritant to US-Indian relations. When, early in 1981, India said it would reprocess used fuel and extract the plutonium, then it was obvious that to continue sending fuel after that would bend the Non-Proliferation Act to breaking point, and the supply was stopped.

Administration officials have worked towards an interpretation of the Act that will allow it to discriminate more in its treatment of customers, between friendly and potentially dangerous countries. The idea is to give a friendly country permission to reprocess used fuel, covering not just one shipment but a continued supply over an extended period, so that the country can make long-term or at any rate medium-term plans. This is in accord with an INFCE recommendation. INFCE said that where the right of prior consent before reprocessing was provided for, 'the criteria for its exercise should be established, to the extent possible, before long-term fuel supply contracts are concluded . . . since such right, if exercised arbitrarily, may have a negative impact upon assurances of fuel supply, and a consequent adverse effect upon national nuclear programmes.'

Chapter 12
THE TIGER IN THE NURSERY

For many people, the only relevant argument about nuclear power is a very simple one: between those who are for it and those who are against it. The question hardly arose in that form when civilian nuclear power was developed. But the intellectual climate has changed in the last two or three decades, and some of these changes are inimical to nuclear power.

There is a new anti-rationalism about, exemplified at one end of the scale by the new interest in astrology and tarot cards, and at the other by the writings of several psychologist-philosophers who elevate the demands of the id over those of the super-ego, such as Norman Brown and Herbert Marcuse. This was a prominent strain in the youth revolts of the late 1960s: 'Down with the reality principle!' was one slogan of that movement.

There is also a new hostility to science, or at any rate to its claims to primacy as a way of exploring and understanding the world. Voices are heard to say the scientific method is not the only way to discover truths, and that there are alternative paths, besides divine revelation. Some writers such as Abraham Maslow and the aforementioned Norman Brown have argued that scientific detachment can be a barrier to understanding. They say the scientific attitude to the natural world is often a subtly aggressive one, using knowledge to attack and conquer an alien nature instead of to better assimilate it joyfully.

At a more homely level, this hostility is merely a widespread suspicion about what scientists are up to as they extend and deploy their powers. A few years ago, a conference was held on the theme of 'The Common Man and the Wonders of Science'. The science writer Nigel Calder drew a parallel with what he saw as the common man's attitude to the wonders of science in an old joke. During the war, some soldiers on a troopship organized a show to entertain the others, and it included a magician. Near the end of his act, the magician said: 'And now, gentlemen, I will perform what I think will be the most amazing, the most astounding trick that most of you have ever seen.' At that moment the troopship struck a mine, and blew up and sank almost immediately. One soldier, clinging to a piece of driftwood, looked around him at the empty sea and said, 'That was a damned silly trick!'

This is manifested also in a concern about the side-effects of technology, compensating (and sometimes over-compensating) for a tendency in the past to overlook these. For instance, in Britain and the United States, the tests that are legally required for a new medicine before it can be put on the market are fifty times as extensive as those required in 1950. The kind of newspaper space that, a generation or two ago, was devoted to stories about new inventions and discoveries is today given over to stories about their dangerous side-effects. In the Press, 'investigative journalism' is all the rage, and many reporters measure their success by the number of dark secrets kept by government departments or big corporations that they can discover. Often this is a valuable public service, but there is a tendency to play up the importance of what has been discovered, and to dramatize or exaggerate its dangers.

There is also a revolt against bigness, against the increasing size of administrative units. The sociologist H. Nowotny wrote in a paper: 'Opposition to nuclear power is . . . directed against "big" industry, seen to be in collusion with "big" government and "big" science. It is the opposition of those who feel powerless and small in the face of these developments.'* Nuclear power is a vast, centrally-directed system of power generation. There is a school of social philosophers and even economists who maintain that small-scale economic units are more efficient and also socially and psychologically healthier.

Liberal attitudes towards national and international politics have become liberal, 'soft' attitudes towards the biosphere and the physical environment. There is a widespread rejection of past chauvinism about the human race and civilization; there is instead a greater respect for 'uncivilized' peoples and for other species, less arrogance and more humility. The traditional way of talking about Man in relation to the environment is the language of subjugation: men 'conquer' the desert and wilderness, and 'tame' rivers and

---

* *Social Aspects of the Nuclear Power Controversy* by H. Nowotny.

jungles. Today these activities are often viewed guiltily, and described in negative terms, as 'plundering' and 'despoiling'.

The roots of this change of attitude are not only subjective; the change stems also from an altered material situation. In past eras, the physical environment was boundless in relation to men's activities, and its resources infinite in relation to his needs. The only limitations were in men's reach. 'As many as the fish in the sea' was a phrase that denoted an infinity. In the present era, Mankind is very much more powerful than it has ever been before, as well as more numerous, and people's activities can have a major impact on the physical environment, even in the short term. People must be restrained in the use of their power because it is so great. The Earth itself is seen today, not as Man's unchanging and unchangeable home, but as a spaceship, with a life support system for its passengers that may prove fragile.

The anxiety about the environment is about aesthetic values as well as dangers. More and more people are coming to feel that they want to live in the natural world rather than a world in which natural things have been vanquished and replaced, or else pushed into distant corners, an organic world rather than a programmed one, that, as the poet E.E. Cummings put it in a beautiful line:

a world of made is not a world of born.

Nuclear energy changes nature more fundamentally than any other technology. It transmutes elements, and creates new elements or new forms of elements. By turning carbon atoms into the isotope carbon 14, nuclear fission has changed the contents of the world's oceans. Strontium 90 did not exist before men split uranium atoms; now, because it is inseparable from calcium in the life cycle, it is present in an infinitesimal quantity in the bones of every baby that is born, a contribution that nuclear fission has made already to the human species, though by test nuclear explosions rather than by nuclear power. It is hardly surprising that some people look upon nuclear fission with anxiety, and wonder whether this particular scientific trick should be performed.

The radiation emitted in the course of nuclear power production worries a lot of people, to the point of turning some into determined opponents of nuclear power. Radiation is a good instance of the way in which every aspect of nuclear power becomes contentious, and the facts are open to many interpretations.

The harm caused by very small amounts of ionizing radiation can only be measured statistically. The International Commission for Radiological Protection calculates that a rem – a measure of received radiation – means 125 cancer deaths per million of the population. Applying this to Britain, one finds that the British nuclear power programme, which supplies 14 per cent of the country's electricity, gives each Briton an average dose of 0.2 millirems a year, or two ten-thousandths of a rem, per year. Most of this radiation comes from the reprocessing of used Magnox fuel at Windscale, not from nuclear reactors. Spread over the population, this means about one extra death from cancer each year. True, this is not the only harm done. The ICRP calculates also that a rem produces fifty cases of genetic disorder per million live births, which comes out at one genetic defect caused by nuclear power every four years or so. It is difficult to fit the pain and horror of the birth of a deformed baby (though not every genetic defect is a serious deformation) into any kind of calculation. Nonetheless, on the basis of these figures, one would have to conclude that nuclear power causes less harm than any other kind.

The impact of statistics such as these often depends on the perspective into which they are put. In May 1979, *Nucleonics Week* reported in a headline that a new study had forecast '0.5 additional cancer deaths per gigawatt/year of nuclear power production in the United States'. In the same week, the *New York Times* reported that a National Academy of Sciences study estimated that 'the price of nuclear power will be the cancer deaths of 2,000 US citizens by the end of the century'. This seems like a much higher price to pay. But both papers were reporting the same study, and both were accurate. The price of nuclear power is put into another perspective by a second National Academy of Sciences report issued at the same time, which said that low-level radiation from all sources would cause 200,000 cancer cases before the end of the century. (Since roughly one-fifth of the population dies of some form of cancer, the total number of cancer deaths in America by the end of the century will be more than 25 million.)

The harmful effects of low-level radiation from other sources makes it easy to point up the absurdity of some of the concerns about the radiation from nuclear power. Natural background radiation is about four hundred times the amount created by nuclear power. Medical use produces two hundred times as much. Colour television produces more radiation, so who's for banning it? Alvin Weinberg, the Director of the Argonne National Laboratory, made a more bizarre comparison. He told a conference that each of us receives 20 millirems (a millirem is a thousandth of a rem) each year from our internal radioactivity, which comes from the potassium-40 that the body contains. Sleeping with a partner in bed gives one an extra 0.3 millirems a year, he said, and this is more than each of us gets from the nuclear power programme. So one could offset the radiation effects of nuclear power by forsaking that pleasure and sleeping alone.

Actually, it has never been established empirically that these very small amounts of radiation do any harm at all. The assumption that they do is made by extrapolating downwards from the measurable harm done by larger doses. The most authoritative American studies of radiation effects are conducted by a committee of the National Academy of Sciences called the BEIR (Biological Effects of Ionising Radiation) Committee. When the BEIR Committee published its last report, in July 1980, with figures that accord closely with the ICRP ones, it noted that a dissenting minority of committee members believe that extrapolating downwards leads to a substantial overestimation of the effects of radiation.

Members of official bodies concerned with the effects of radiation all believe that their estimates are conservative, so that if they err at all it is on the side of caution. However, there are a few critics who claim that they underestimate the effects. The best-known is Thomas Mancuso, a University of Pittsburgh specialist in radiation medicine. He directed a study for the AEC that was discontinued, whereupon he claimed that the AEC stopped it because it did not like what he was finding out, and went into public opposition with charges that the AEC was covering up radiation dangers. Two other scientists, Arthur Tamplin and John W. Gofman, have argued that the toxicity of plutonium is much greater than official accounts say, because of the way it collects in the body; this will be an important issue if breeder reactors become widespread. (By now, atomic energy officials are so used to countering arguments by Gofman and Tamplin that they refer to the pair simply as 'G and T', or 'gin and tonic'.)

Most studies of the effects of plutonium appear to contradict this view, including those by Britain's Medical Research Council. So does the survival of two groups of people who were accidentally exposed to alpha radiation in American laboratories many years ago; if the Gofman-Tamplin figures were correct, some of them should be dead by now. Arguments in this area are often accompanied by accusations of dishonesty on both sides. Those who dispute the official view are a small minority of scientists, but there is enough argument to suggest to members of the public that the question of radiation effects is an open one.

Certainly in the past, radiation was treated with less respect than it deserves. Precautions in the use of x-ray machines that are routine today have only been adopted in the past thirty years; among people who worked with them before that, there are more cancer deaths than among the general population. In the nuclear age, we have seen how the uranium mining industry ignored the radiation from radon gas. Nuclear weapons tests were staged in Nevada in the 1950s without sufficient regard to radioactive fallout, so that the cancer and leukaemia rate among servicemen who participated and among nearby residents is higher than the national average. In one test, the rate of leukaemia among the two thousand GIs who observed it was found twenty years later to be more than twice the national average. Even so, the incidence was not dramatic: there were eight cases, against a statistical norm of 3.5.

The fraction in the figure of 3.5 leukaemia cases points to the abstract nature of deaths caused by low-level radiation. They are not abstract to the people who fall ill and die. But no one can point to one of those ex-servicemen who contracted leukaemia and say that he is the victim of a bomb test. The deaths are manifested in an increase in the number that would occur anyway.

Discussion of this area becomes more abstract still when one is talking about the big nuclear power accident that might or might not happen one day. Here, one is talking about a very small possibility of a large number of deaths, and translating this into everyday risks. For instance, one study concluded that the worst possible nuclear power plant accident in America could result in 3,400 deaths. But, it said, this was so unlikely that statistically, if there were one hundred large nuclear power plants operating in the United States, each person's chance of being killed in a nuclear power accident in any one year was less than one in 50 million. (This study, by the Brookhaven Laboratory, was later discounted on the ground of its methodology, so this figure may not be accurate, but the principle remains.)

Behind statistics such as these lies the technical discussion of whether a big accident involving thousands of deaths could actually occur, like the 'China syndrome', made famous by the film of that name. This is the most dangerous reactor accident: the coolant in a reactor, either water or gas, is dissipated somehow and the emergency core cooling system designed to replace it in such an emergency does not function properly: the reactor core over-heats, like a saucepan that has boiled dry. The molten metal burns through the floor, and goes on downwards, 'in the direction of China' – hence, the 'China syndrome'. Few scientists accept that even in this unlikely eventuality, the radioactive material spreading out into the earth would make an area the size of Pennsylvania uninhabitable, as someone in the film said.

The technical argument over the safety of nuclear power is beyond the scope of this book. However, doubts about safety are a major factor in generating opposition to nuclear power, and therefor affect powerfully both its politics and its economics.

There have been leakages of radioactive materials, most notably from the Windscale plant in Britain in 1957 (and on other occasions, to a lesser degree), from the storage tanks at Hanford, Washington, at the La Hague reprocessing plant, and at Three Mile Island. It is believed that the only deaths directly attributable to radioactivity at a nuclear power plant occurred in Czechoslovakia's one nuclear power plant, when two workmen ignored safety precautions and made a mistake as well; even this accident was never announced, but its existence and nature have been inferred by Western scientists reading the technical literature. A few defectors from the American nuclear power industry have joined the anti-nuclear camp and say that a serious accident is very possible. Most people inside the industry and, for that matter, a majority of scientists outside it believe that the sequence of events that could lead to one is wildly improbable. One cannot go further than that in any situation where human error can be a factor. No one looking ahead to conceivable accidents at a nuclear power plant would be likely to include the possibility that a candle flame would set fire to vital cables, but this is what led to the accident at Brown's Ferry nuclear power station in Alabama in 1975 that shut down the plant for months.

It is unfortunate that public discussion of nuclear safety always takes place in an atmosphere of adversary proceedings, so that every statement and every statistic may be put to the service of an argument. As Professor Harvey Brooks, the Harvard physicist, said once, 'What drives me up the wall is that conservative expressions of scientific caution are seized upon by nuclear critics as admissions that nuclear power is unsafe.'* This kind of atmosphere is conducive neither to rational discussion nor to optimal safety measures.

The nuclear industry is watched more closely and fearfully, and judged by far stricter standards, than any other. The spotlight is thrown on every flaw. As this was being written, a chemical factory caught fire in Rainham, Essex, and because there seemed to be a possibility that toxic gas might escape, a thousand people living nearby were quickly moved out of their homes. In the event, no harmful gas was released and the people returned to their homes the following day. The event received a bare mention in a few newspapers. If, however, the fire had been at a nuclear power facility of some kind, and homes had been evacuated because of the possibility of a release of radioactive gas, this would have been a major international news story. Television teams from a dozen countries would have descended on the area, and Rainham would have become as famous as Harrisburg. Incidentally, this prospect might mean that in the case of an accident at a nuclear facility, the managers might draw back from recommending precautionary evacuation.

During the same year, 123 men were drowned in the collapse of an oil rig in the North Sea, and forty-nine children were killed when a propane gas tank exploded at a village school in Spain. No one suggested, following these accidents, that the exploitation of North Sea oil should be curtailed, or that propane gas should never again be used for heating purposes. But if an accident at a nuclear power plant caused tragedy on this scale, it would lead to worldwide demands that nuclear energy be dispensed with as a power source,

* Quoted in *Energy Futures*, by Robert Stobaugh and Daniel Yergin.

presumably (though this would not be said) in favour of others such as oil and propane gas.

Yet to some people, the power to poison contained in nuclear fission is different in kind from these others, so different that it cannot be measured on the same scale, unique in the nature of its horror, pervasive through space and time, continuing down from generation to generation, so that no safety measures that might be incorporated into its use are sufficient. It is like chaining a tiger in a baby's nursery, and asking the baby's mother to accept it. She will not. So far as she is concerned, the chain holding that tiger will never be strong enough.

* * *

The comprehensive case against nuclear power does not rest principally on safety considerations, but is based on social values and economic analysis. In this view, a high-technology, capital-intensive form of power generation is the wrong way to tackle the energy problem; and in fact, this problem is not properly understood if it is is seen simply as a need to produce more energy, without any regard to different ways of using energy and their social costs and consequences.

This view is exemplified in the writings of Amory Lovins, a physicist and economist who has summed up his case against nuclear power in his book *Soft Energy Paths* (again that emotionally loaded adjective 'soft').

Giving a resume of his beliefs, Lovins writes in this book:

> The energy problem should not be how to expand supplies to meet the postulated extrapolative needs of a dynamic economy, but rather how to accomplish social goals elegantly with a minimum of energy and effort, meanwhile taking care to preserve a social fabric that not only tolerates but encourages diverse values and life styles. . . . The technical, economic and social problems of fission technology are so intractable, and technical efforts to palliate these problems are politically so dangerous, that we should abandon the technology with all deliberate speed.

To Lovins as to many of the anti-nuclear protesters, his argument against nuclear power is almost incidental to a larger world-view. In this view, a high technology, centralized power source contains all the social and economic evils of bigness. As he writes:

> In an electrical world, your lifeline comes not from an understanding neighbourhood technology run by people you know who are at your own social level, but rather from an alien, remote and perhaps humiliatingly uncontrollable technology run by a faraway, bureaucratized, technical elite who have probably never heard of you. Decisions about who should have how much energy at what price also becomes centralized – a politically dangerous trend because it divides those who use energy from those who supply and regulate it. Those who do not like the decisions can simply be disconnected.

He worries about biological as well as social alienation:

> Life near enough to the soil to understand its rhythms, balances and tensions has nurtured every previous culture, and for all its hardships in the past – most of them now avoidable – it has a cultural value; are we sure that by rejecting that awareness of the life process and turning ourselves into mere cogs of a biotic production, we are not losing something essential to the human psyche and to our own mythic coherence?

It is essential to the arguments of the anti-nuclear campaigners that soft energy sources are feasible: they do not advocate a return to primitive life-styles. This means, in Lovins' words, technologies

> that rely on renewable energy flows that are always there whether we use them or not, such as sun and wind and vegetation: on energy income, not on depletable energy capital . . . flexible and easy to understand and use without esoteric skills, accessible rather than arcane.

It must be matched in scale and geographical location to the end use of the energy. These technologies include solar, wind and wave power, and the production of alcohol for fuel from plants (as is already practised on a large scale in Brazil – see Chapter 5). Conservation, and more efficient use of energy by consumers, also plays a big part in these hopes for the future.

These views are challenged, of course, in particular the figures that Lovins produced to show that soft energy paths are feasible.

Many of his opponents see the soft energy path as chimerical – 'the yellow brick road', as one group called it in a collection of papers brought out specifically to rebut his ideas. Certainly some calculations regarding alternative technologies seem like wishful thinking by people who want to escape from some of the disagreeable aspects of modern industrial society while preserving its comforts. Intelligent people argue on both sides. The author of *Small Is Beautiful*, the economist E.F. Schumacher, was working on a sequel when he died in 1977 to be called *Small Is Possible*.

Many people involved in nuclear power simply cannot understand how any intelligent person can be anti-nuclear. As they see it, the world faces a critical energy shortage that threatens everyone's standard of living. Nuclear power was discovered in the nick of time as an energy source that compares favorably economically with any other and has proved to be far safer. How can any person of goodwill reject it? These see a price being paid for every delay; so many billion barrels of imported oil for a stretched-out reactor licensing procedure, or for a moratorium on fast breeders. The antis, on the other hand, see the nuclear power advocates as technocrats with closed minds, hooked on the most shallow values of a consumer society and on the ingenuity of high technology, who deceive others and perhaps themselves about economics and safety in order to push Mankind further down the wrong road and poison the planet, in defiance of common sense and human values.

Opponents have contributed to the slowing down of the expansion of nuclear power. They have employed all the legal and administrative devices available to erect new hurdles to the construction of any new nuclear facility, with the help of expert critics to ask the right technical questions. In some states, a utility wanting to construct a nuclear power plant has to satisfy nine different regulatory bodies. Five nuclear power plants were shut down in 1979 because it was discovered that a mistake by a computer operator had led to an error in calculating the geological stability of the sites.

All this means that the average construction time for a nuclear power plant in America has been increased to twelve years, compared with four in Japan. Since construction is the most expensive part of the operation of a nuclear power plant, and since interest payments are a large part of the cost, this stretching out has increased enormously the cost of a nuclear power plant, while the uncertainty has reduced its appeal. The Deputy Secretary for Energy under President Carter, John O'Leary, went so far as to say in a television programme: 'Nuclear power is finished as a short-run proposition in this country. The utility industry simply isn't going to buy new plants for the foreseeable future. . . . It is just too hard for a utility to go through the long, long process of licensing, and at the end of it not be permitted to run the plant.'*

The organized anti-nuclear movement in America is very much a carry-over from the protest movement against the Vietnam War, engaging many of the same emotions and, indeed, many of the same personalities. Most of the participants are middle-class liberals, with a strong college faculty representation, particularly in local campaigns against a nuclear facility. † The enemy is big business, the big corporations that mine and refine uranium, others that build nuclear reactors, and others that produce electricity from them, they and their lobbyists and public relations men. Activists in the anti-nuclear movement have no doubt about the industry's ability and will to cover up dangers to the public, and not much doubt about who killed Karen Silkwood, whose car crashed while she was on her way to tell a reporter about slip-ups and cover-ups at a Kerr McGee uranium mill.

In most other countries, atomic power is wholly or partly Government-operated. In France, protests have often led to clashes with police. This is partly because actual demonstrations are usually directed, not against nuclear power in general, but against the erection of a particular nuclear facility in a particular place; partly because they are often in areas where there is a strong sense of regional identity, and resentment at the control exercised from Paris, particularly Alsace and Brittany; and partly because of a sense of frustration at the Government's determination to steamroller ahead with its nuclear power plans. The Government has made

* NBC's *Meet the Press*, 12 March 1979.

† It is by no means axiomatic that opponents of nuclear power are leftwing. Reactionaries also swim against the current. Some rural conservationists are simply fighting social change, and in Italy the neo-fascist MSI (Italian Social Movement) has a network of youth camps where young people exalt the pastoral life, enact pantheistic rituals, and oppose nuclear power.

nominal concessions. It passed a Freedom of Information Act along US lines in 1976, environmental impact reports are now required, and there are local inquiries into any plans for new nuclear facilities. But civil servants make it almost impossible to get much information through the Freedom of Information Act, environmental impact reports are usually insubstantial, and most people have little confidence that the outcome of a local inquiry will affect the Government's intentions. The second largest trade union group, the CDFT, came out against the expansion of nuclear power, and drew attention on behalf of its members who worked at nuclear power facilities to radiation leaks. This was a factor in swinging much of the Socialist Party over to a position of sceptical agnosticism on nuclear power, an attitude embodied in the Mitterand Government's policies.

In Britain, where civilian nuclear power began soonest and is extensive, there has been curiously little opposition. The issue is argued out from time to time at a public inquiry into a new project, in letters in serious newspapers, and in the occasional television debate. There are ecologically-minded groups that hold meetings and publish newsletters. But it is not a cause that draws in thousands and tens of thousands. It may be that this reflects a condition of psychological depression in the British people, which produces lassitude, and saps away the energy and hopefulness that are required to try to change things.

In West Germany, however, opponents have used the courts successfully to slow the development of nuclear power. There is strong opposition to the fast breeder programme in the Social Democrat and Free Democrat parties, though the Social Democrat Chancellor Helmut Schmidt is committed to it. There is also an anti-nuclear Green Party, which won a handful of votes in the 1980 election. The electricity companies are private but with state government representation on their boards. Private industry, governments and even the universities, through industry-financed research projects, interlock at every stage of nuclear power, so that opponents feel that they are challenging the Establishment, and occasionally risking their careers by doing so. (One reason that some Austrians give for the 'no' victory in the Austrian referendum on nuclear power in 1979 is that nuclear power in that country is contained entirely within Government organizations, and has no private pressure groups.)

The argument crystallized over the plan to build a nuclear fuel reprocessing plant and underground waste storage site at Gorleben, in Lower Saxony, only a few miles from the East German border. The salt cavern deep under the surface seems to make it ideally suited for storing radioactive waste material. Faced with fierce opposition to the plan, the Premier of Lower Saxony, Ernst Albrecht, convened an international panel of experts at Hanover, the state capital, under the chairmanship of Carl Von Weiszäcker, the doyen of German physicists. Some 35,000 anti-nuclear demonstrators converged on Hanover to make *their* views known to the panel. They found an unexpected focus for their protests.

Von Weiszäcker banged his gavel and called the meeting to order in the hangar-like hall at nine o'clock on the morning of Wednesday, 28 March, which was four o'clock in the morning in Harrisburg, Pennsylvania. There, at almost exactly that moment, the alarm bells were sounding in the nuclear power plant at Three Mile Island to indicate that something was wrong. Over the next few days, the news from Three Mile Island formed a background chorus to the testimony given at the hearings. The panelists brought in transistor radios so that they could listen to the latest reports during coffee and lunch breaks. One of the participants was Gene Rockland, a Californian physicist who had done a lot of work on accident analysis; he learned from a telephone conversation with his wife at home that the problem was a hydrogen bubble and a drastic rise in temperature, and he came in and told the others: 'For Christ's sake, you guys, it's really happened!' In the evening, they saw a map of Pennsylvania on German television screens with a large area labelled '*Dackuirung*' – 'Evacuation' – an exaggeration, as it turned out. On the streets outside, demonstrators carried hastily-painted banners saying 'No Harrisburgs Here' and 'We All Live In Pennsylvania'.

The opposition won in Hanover: the Gorleban project was postponed indefinitely. State Premier Albrecht made the decision to do this after the Social Democrat opposition in the state Parliament came out against it, in the middle of the panel hearings. Announcing the postponement, he made it clear that public opinion was a major factor in his decision. He said that even if the project was technically feasible, two questions remained: 'Whether construction of such a facility is indispensible, and whether it can be carried through politically.'

Von Weiszäcker says today that the Three Mile Island accident convinced him that nuclear power is safe; so many things went wrong, and yet no one was killed. But he also says he has swung over to an anti-nuclear position, at least to the extent of believing that Germany can and should get along without nuclear power. He was persuaded of this, he says, by reading Amory Lovins.

* * *

The one country where the argument over nuclear power was so intense that it became a major election issue is a country that does not have any nuclear power: Australia.

The debate was anything but academic. It was over whether Australia should mine uranium. Since Australia possesses a fifth of the known uranium deposits in the non-Communist world, and far more than a fifth of the uranium available for export, the outcome was very important.

Much of it was remarkably disinterested. Australians argued passionately about whether nuclear power plants were safe, about the risks of nuclear terrorism and proliferation, and whether waste can be disposed of safely. But they were arguing about the dangers to other countries, not themselves. Australia, with its small population and vast expanse, has so much easily accessible coal and natural gas that nuclear power has no economic attraction, and there is no move afoot to build nuclear power plants. The argument is mostly about the morality of helping to build up nuclear power elsewhere, and exporting dangers, though there has also been some concern about mining damaging the environment.

Australian uranium mining all but closed down during the market slump of the early 1970s; but during this period, the Government encouraged exploration with subsidies, and this produced several big new discoveries, mostly in the Northern Territories. These were made by consortia in which foreign companies had a major share: Exxon, British Petroleum, Urangesellschaft and Minatome, among others.

Australians were drawn into the worldwide argument over

nuclear power. More and more, its morality and its environmental effects were discussed and debated. Many Australians who had earlier campaigned against the anti-Communist chauvinism that accompanied Australian participation in the Vietnam War now turned to fighting what they saw as unthinking technology. They made much of the fact that they were arguing, not merely against big business, but against foreign big business. The opposition included most of the Labour Party and most of the big trade unions.

Gough Whitlam, the Labour Party Prime Minister who was elected at the end of 1972, ordered a moratorium on uranium mining. The argument continued, with debates in Parliament and demonstrations in the streets.

One factor is that the uranium is located mostly in the vast open spaces that are the home of the Aborigines. The involvement of the Aborigines touched a nerve in many Australians, and seemed to exemplify the clash of values that lies at the heart of many passionate arguments about nuclear power. Traditionally, most Australians have regarded Aborigines with cheerful and not always unfriendly contempt. Aborigines do not take naturally to the work ethic and the attributes that have built up Australian society; certainly when they move into the cities and try to adopt Western ways, they often lapse into drunkenness and squalor, like American Indians. But in their own environment, they seem to embody in their lives the social, biological and spiritual virtues that are the antithesis of fragmented, industrialized society, and which seem to some to merit respect and humility. Many of the lands which are being explored for uranium contain some of the finest Aborigine rock paintings, and are regarded by the Aborigines as sacred. The heavy metallic tread of technology in this landscape, with the ravaging of its surface – for most uranium mining in Australia is open-cast strip mining – seemed to some to be an abomination. The Australian novelist Patrick White referred to his country becoming 'increasingly scarred morally as well as physically' by uranium mining.*

* In a letter to Mary Elliott of the Australian branch of Friends of the Earth. That this conflict and this feeling is widespread, and touches on some of the universal themes of today cited at the beginning of this chapter, is seen in the following passage in a protest leaflet: 'The people of Carswell Lake . . . are being

The Whitlam Government fell, and the argument went on. In 1974, its successor, under Prime Minister Malcolm Fraser, appointed a commission to study the environmental aspects of uranium mining in what was known as the Ranger Field in Northern Australia, under a Supreme Court Justice, William Fox. Justice Fox turned this into a wide-ranging study of nuclear power and its morality. The question: 'Should uranium be mined in Northern Australia?' also meant, so far as he was concerned, 'Are nuclear reactors safe?' and 'How easily can the plutonium they produce be made into nuclear bombs?' The Government expected the commission to complete their report in nine months. In fact, it took two years, and the final report, which runs to six hundred pages, covers every aspect of nuclear power.

Like the US Government at this time, the commission put nuclear weapons proliferation at the centre of the issue. The report says: 'The nuclear power industry is unintentionally contributing to an increased risk of nuclear war. This is the most serious hazard associated with the industry.'

It also shows a sensitivity to the needs and the rights of the Aborigines that has not often been found in the past in official pronouncements. The report says: 'Of all the relationships that traditional Aboriginal Man has with anybody or anything, the most important is that which binds him to a particular tract of land, which he refers to as "his country". This is a religious bond. The people have a spiritual relationship with the species and physical features of their environment.'

Despite this, it decided that, with a combination of safeguards and compensation for the Aborigines, uranium mining on much of this land should go ahead: 'That our values are different is not to be

---

forced against their will to give up their ancestral lands and sacrifice their means of livelihood for the sake of "economic development." Abrogating its legal responsibility to consult with native people on the proper allocation of lands and resources which are theirs by Aboriginal title and treaties, the Government has given a multinational corporation full permission to develop one of the richest known uranium deposits in the world, near Carswell Lake.' From the wording, this could have been issued in Australia, but Carswell Lake is in northern Saskatchewan, the Aboriginal people referred to are the Dene Indians, and the leaflet was issued by a Canadian citizens' group.

denied, but we have nevertheless striven to understand as well as can be done their values and their viewpoint. We have given careful attention to all that has been put before us by them or on their behalf. In the end, we form the conclusion that their opposition should not be allowed to prevail.' It recommended the establishment of a huge protected area near the mining sites, and other measures to limit the impact of mining.

Like the Parker Report on Windscale in Britain a little while later, it brought in Article IV of the Non-Proliferation Treaty, and said that a refusal to export any uranium at all would place Australia, as an NPT signatory, in breach of the treaty, since Article IV obliges signatories to facilitate the transfer of nuclear technology.

The Ranger Report was issued in May 1977. Shortly after this, Prime Minister Fraser visited Washington. There, President Carter asked him to release Australian uranium for export to Europe and Japan to reduce the incentive to go in for fast breeder reactors.

So in August, Fraser announced that the mining and export of uranium would be resumed. He told Parliament about it in a speech that matched the loftiest ethical tones of the opponents of uranium mining: 'The decision has been motivated by a high sense of moral responsibility to all Australians and to the community of nations,' he said. He barely mentioned commercial benefits. The recommendations of the Ranger Report were followed almost to the letter. The strict anti-proliferation controls on uranium exports included are substantially the same as the ones America and Canada impose on their sale of nuclear fuel: only to NPT signatories, and no enrichment, reprocessing or transfer without prior consent.

As mining teams moved back into the Northern Territory, opposition grew in intensity. The Australian Council of Trade Unions banned work on any new mining sites. An election was called for December 1977, and uranium mining was a key issue. The Labour Party attacked Fraser's decision, and even warned potential purchasers of Australian uranium that if it came to power, it would repudiate contracts entered into by the Fraser Government. But after Fraser and the Conservatives won the election, some of the heat went out of the issue, and it did not figure prominently in the next election, in October 1980.

The results of all this effort were initially disappointing. The expected uranium boom has not taken place. The mining companies did not find the world market in uranium encouraging. Though they spent a lot of money campaigning for the right to mine uranium, they are not now rushing to do so. The Government has had to relax the limit placed on foreign involvement in order to attract the capital to start new ventures. However, there is now some demand for Australian uranium. The Australian Government painstakingly negotiated an agreement to export it to Euratom. Every Euratom government sent a representative to Canberra in the course of the negotiations, which centred on precisely what the Euratom countries might do with the uranium. Eventually, Australia agreed to the free transfer of fuel among Euratom countries, but would not allow its transfer outside the European Community without special permission. Euratom countries plan to send some of this uranium to the Soviet Union to be enriched. Japan is also interested in buying some.

Chapter 13

BREAKING THE RULES

In November 1968, the Belgian company Société Générale des Minerais sold 200 tons of uranium oxide, or yellowcake, to a German chemical firm, Asmara Chemie of Wiesbaden. The yellowcake was refined from an unsold stockpile of uranium from the Shinkalobwe mine in the former Belgian Congo, now Zaire: SGM is a part of the same conglomerate as the Union Minière de Haute Katanga. Asmara Chemie had told SGM that it wanted to use the uranium oxide as a catalyst in the mass production of petrochemicals, and it explained this in terms that sounded plausible to SGM, who in any case were anxious to get rid of their stockpile. Asmara Chemie also said it was going to send the yellowcake to a chemical processing company in Milan, where it would be turned into the form suitable for its use as a catalyst. Since the movement of the uranium would be within the European Community, no special Euratom permission was needed.

The 200 tons of uranium oxide powder, packed in steel drums, was sent by train from the SGM warehouse to Antwerp, and there loaded on the freighter Scheersberg A, which was to take it to Genoa. The ship sailed out of Antwerp, but did not arrive in Genoa. The Scheersberg A vanished, along with those steel drums of yellowcake. Euratom investigators were to find later that Asmara Chemie was a very small company which lacked the resources to use or even store 200 tons of uranium oxide, and the Milan chemical company concerned had no facilities for treating the substance. Nonetheless, SGM was paid for the uranium oxide, just over a million dollars; the money had been lodged in a Swiss bank before the uranium was transferred.

We know now that the uranium went to Israel, and that the complicated deception involving fake sales was worked out by the Israeli secret service, Mossad. The story came out five years later, when a Danish Mossad agent, Dan Aerbel, was arrested in Norway following a Mossad killing of a Moroccan. Aerbel told Norwegian police all about the Scheersberg A operation, in which he had played a part. Later, back in Israel, Aerbel denied that he ever made the confession. Nevertheless, the details tie in with everything known about the disappearance of the uranium. The whole story was brought out in a book, *The Plumbat Affair*,* 'Plumbat' being the code name that Mossad gave to the operation, according to Aerbel.

The only mystery remaining is why Israel went to so much trouble to acquire natural uranium. It is true that its supply from France had been stopped. But uranium can be bought from a number of countries that might not impose political conditions, and small quantities have been mined within Israel, from the potash deposits near the Dead Sea. Perhaps the Israeli Government believed that a purchase of uranium from South Africa could not be kept secret for long, and might give rise to questions about what Israel was doing with it; perhaps it yielded to a Mossad penchant for covert operations.

There are a number of reasons for believing that, despite repeated Israeli denials, it has armed itself either with atomic bombs or with material and components that can be made into one very quickly. The facts of the Scheersberg A shipment constitute the most persuasive reason of all. Israel could have bought the uranium openly if it wanted to use it only for a peaceful purpose. Instead, it went to great trouble to acquire it by subterfuge. If someone tunnels into a bank, there is strong reason to believe that he intends to rob it: if he wanted simply to withdraw some money from his account, he would presumably have entered by the front door.

The episode of the Scheersberg A and its uranium cargo draws attention to the loopholes in the régime governing trade in nuclear materials. The more common and the more widespread the materials to be controlled become, the more likely there are to be gaps in the controls.

When nuclear proliferation is discussed, lists are often made of near-nuclear weapons nations. There could also be a list of possible nuclear weapons nations. Israel would be a strong candidate to head this list, because of the combination of capability and motivation.

Israel has been involved in nuclear technology for a long time, as

---

* *The Plumbat Affair*, by Elaine Davenport, Paul Eddy and Peter Gillman

it has in almost every other advanced technology. Israeli scientists patented a method of separating plutonium, and one for producing heavy water, both of which have been used experimentally in France. Israel bought a small research reactor from America under the atoms-for-peace programme and highly enriched uranium to fuel it, all under safeguards. It made plans for a nuclear-powered water desalinating plant, but then enthusiasm waned.

It also embarked in secret on a more far-reaching programme of nuclear co-operation with France. The agreement with France was signed during the most torrid phase of the Israeli-French friendship, during the months of the Suez Canal crisis in 1956 which culminated in the Anglo-French-Israeli attack on Egypt. The Israelis built a reactor with French help at Dimona, an isolated spot in the Negev Desert, and told the world that it was a textile plant. There was also a plan for France to help Israel build a reprocessing plant, but this was scotched by General de Gaulle when he came to power and tilted French policy gradually in favour of the Arabs.

US Intelligence discovered what was being built at Dimona in 1960, and President Eisenhower was furious at Israel for the deception. Israel promised that it had no intention of producing nuclear weapons, but it has always resisted pressure to sign the NPT or accept safeguards on Dimona. French officials say that any nuclear aid they have given including Dimona is under strict French safeguards. But the terms of the agreement on Dimona have never been made public, and it is difficult to believe that France is able to tell Israel today what it may and may not do with a reactor. The US Administration believes that Israel has built by itself a small reprocessing plant in which plutonium can be extracted.

There have been many reports in recent years that Israel has constructed atomic bombs. In fact, there seems to be an attempt in certain quarters to persuade the world of this. A 1974 CIA report on nuclear proliferation which said Israel had probably built a bomb was made public: then the CIA said this report should have remained secret, but was declassified by mistake. Later, the CIA investigated the disappearance of a quantity of highly enriched uranium from the Nuclear Materials and Equipment Company plant at Apollo, Pennsylvania in 1965, and concluded that it had been taken to Israel. However, the Nuclear Regulatory Commission decided that the material was not stolen or diverted, but was lost over a period of years in an industrial process.

* * *

Pakistan, like Israel, says it intends to use nuclear energy for peaceful purposes only, and like Israel, it has engaged in a complicated and costly enterprise to circumvent the controls over weapons-related materials.

After the Indian explosion, it was obvious that Pakistan might want to follow suit, whether for reasons of prestige or of military security. Prime Minister Ali Bhutto said as much in a much-quoted interview soon after the Indian explosion, promising that Pakistan would have an atomic bomb 'even if we have to eat grass'. This could have been just an outburst of emotion rather than a statement of policy. However, the BBC television programme *Panorama* reports Pakistani diplomats as saying that Bhutto's decision to build a nuclear bomb was taken even before the Indian explosion, in 1972. Certainly the reprocessing plant that Pakistan wanted to purchase from France would have made no sense economically as part of Pakistan's civil programme.

According to the informants of the *Panorama* programme – and there were more than one – Bhutto started negotiations with the Libyan leader Colonal Gadaffi which led to Gadaffi supplying Pakistan with large sums of money to help it with its bomb project. What was promised in return is unclear; Gadaffi apparently wanted a nuclear bomb, but was promised only access to all the technology. Libya also supplied uranium to Pakistan. There are newspaper reports that Libyan troops entered the Republic of Chad, on the southern border, and hijacked some truckloads of uranium travelling overland from Niger. If this happened it would have been an unnecessary adventure. Libya already had uranium to spare; it imported 258 tons from Niger in 1978 and '79, and has no nuclear energy project in which to use it.

The Pakistan Government found a use for the Libyan money. It set out to build a centrifuge enrichment plant, buying the components in Europe. A key element in this scheme was industrial espionage. A Pakistani metallurgist, Abdel Khan, took a job in Almelo, Holland, with a Dutch firm doing contract work for the Urenco enrichment plant in Almelo, and for a while in the centrifuge company itself. Speaking several European languages and technically qualified, he worked as a translator, and saw many documents. He obtained a list of all Urenco's sub-contractors.

Then Pakistan set about ordering the components for a plant similar to the Urenco one from the sub-contractors. It feared that it might not be allowed to buy them openly; Pakistan's only reactor is a CANDU, which uses natural uranium, and so Pakistan has no need for enrichment for its civil programme. It set up a chain of dummy companies in several European countries and ordered inverters in Britain, aluminium parts and motors from firms in Germany, valves and tubing from firms in Switzerland, and other tubes for the rotors in Holland. All the firms approached were Urenco suppliers.

Inverters are devices to regulate the flow of electrical current. These were ordered from Emerson Electrical Industrial Controls in Swindon, to be delivered to Weargate Ltd, in Swansea. Emerson were told that they were to be used in a textile plant. But someone at Emerson noted that the specifications were far more exacting than those needed for a textile plant, and were similar to those for the inverters that Emerson made for Urenco. He told the Department of Trade, but no action was taken. So he told a friend of his who knew Frank Allaun, a member of parliament and Campaign for Nuclear Disarmament activist, and suggested that Allaun raise the matter in Parliament. Allaun did, with a question to the Secretary for Trade, Edmund Dell, asking whether approval was given for the supply of equipment which could contribute to Pakistan's capability to build an atomic bomb. Dell replied that no approval was needed because the inverters did not come under export controls. Allaun pressed ahead, both with more questions in the House and with private representations. The Department of Trade put export controls on inverters two months after Allaun's original question. But two-thirds of the number ordered had already gone to Pakistan.

An international intelligence operation was mounted to trace all the other purchases that were made for the enrichment plant. Most of the items that Pakistan bought were standard industrial equipment and were not specific to a centrifuge enrichment plant (though two Dutch companies broke the law in evading export regulations). Four governments agreed to halt the sale of most of the relevant equipment. The Swiss Government refused to do so on some items, contending that these were normal commercial transactions and no specifically banned items were involved. American officials in particular told Switzerland it was breaking the spirit of the NPT, if not the letter. The United States also wielded the stick it held as a nuclear supplier, to the anger of the Swiss; it would not give Switzerland permission to send some used fuel of US origin to Cap la Hague for reprocessing. This quarrel was patched up. The Swiss agreed to look more closely at exports, and the United States gave permission for reprocessing.

The United States also remonstrated with Pakistan, to no avail. The Pakistan Government continued to deny that it was working towards a nuclear explosive. The US Government set up a mission consisting of members of the Arms Control and Disarmament Agency and the Nuclear Regulatory Commission to discuss the matter with the authorities in Islamabad, but the Pakistan Government refused to receive them. US military and economic aid to Pakistan was halted, under the Symington Amendment. Canada also halted the supply of uranium and heavy water for the CANDU. When the Soviet invasion of Afghanistan brought Soviet forces to the Pakistan border, the Carter Administration was ready to rush military aid to Pakistan, but Pakistan turned down the offer. It made it clear that this was because the US Government had wanted to impose conditions concerning its nuclear power programme. Under Reagan, military aid was sent anyway.

The covert purchase of equipment is continuing. Several Middle Eastern companies have bought quantities of expensive measuring instruments, ostensibly for use in their own countries, but actually for trans-shipment to Pakistan, and more inverters have been bought in Canada. Western governments cannot always stop this, and sometimes the trail of intermediaries is so complicated that they cannot even detect it.

Pakistan seems also to be setting out on the other route to the bomb, the plutonium route, perhaps because construction of enrichment centrifuges is proving slower and more difficult than was expected. Before the French contract to supply a reprocessing plant was cancelled, blueprints for the plant were delivered to the Pakistan Atomic Energy Commission. It set out to buy some components for this abroad, some from an Italian company that is owned by the French company, BSL, which was originally to build the reprocessing plant.

Evidently, it is possible to block trade in fissile materials, at least from most countries able to supply them. It is much more difficult, if not impossible, to prevent a country from developing nuclear technology by itself, even if it has to purchase some items from abroad.

* * *

Iraq's nuclear facilities, and the questions surrounding their capability and purpose, were brought to the world's attention explosively on 8 June 1981, when Israeli fighter-bombers suddenly appeared in the sky over Bagdad, and swooped down to blow a nuclear reactor to pieces with 16 tons of bombs. This was an unprecedented act of unilateral anti-proliferation policy, with many implications.

The raid was preceded by a number of expressions of anxiety about Iraq's nuclear research programme over three years. Yet curiously, it did not follow from them, but branched off in a different direction. This was scarcely noted in the flood of comment that followed the raid.

Iraq began a stepped-up programme of nuclear research in the 1970s. It had already acquired a small research reactor from the Soviet Union which was sited at Taiwatha, near Bagdad. It sent hundreds of young scientists abroad to study nuclear physics and engineering in several European countries and the United States, and set out to expand its research facilities at Taiwatha. In November 1975, the Iraqi Government signed an agreement with France to buy from France two large research reactors, which the French named Osirak and Isis, and also the fuel for them, which was 93 per cent enriched uranium.

The US Government made the same kind of representations to France over the Iraqi deal that it had made over the one with Pakistan. It was worried, as others were, by the prospect of Iraq acquiring a supply of 93 per cent enriched uranium, which was bomb grade. However, it was more difficult for France to pull out of this deal than the one with Pakistan. For one thing, Iraq had signed the NPT and agreed to full-scope safeguards. For another, Iraq supplied 25 per cent of France's total oil consumption, at least until the war with Iran curtailed exports from Iraq, and offending it could be costly.

There were signs that the French Government was having second thoughts, once again. It asked Iraq to accept a lower-enriched fuel element called 'caramel' (see next chapter) which it said would serve just as well for research purposes, but the Iraqi Government refused. And after the first half-built Osirak reactor was destroyed by plastic explosives in a factory near Toulon in April 1979 – it is widely assumed that Mossad, the Israeli secret service, was responsible – one man who had many dealings with French nuclear energy officials in the following months said that none of them looked particularly unhappy when they talked about the episode.

Italy also played a part in Iraq's nuclear power programme. It sold Iraq a 'hot cell', a low pressure chamber in which small amounts of plutonium can be extracted and handled safely. Working with this is essential in learning how to handle and work with plutonium. The Italian Government also bought 10 tons of uranium from a German nuclear materials brokerage firm and sold it to Iraq. German officials indicated that they would have held up the export licence if they had known the uranium was destined for Iraq. (Iraq also bought some uranium from Portugal and Niger.) Italy has both strategic and commercial motives for selling nuclear technology to Iraq. The Italian Government wants to maintain good relations with a major oil supplier. Also, the companies which produced Iraq's hot cell had lost orders through the cutbacks in Italy's nuclear power programme, and like other European companies in the field, they want to make up for the loss of domestic orders by exporting. The Government is behind them because it wants to keep the nuclear energy industry healthy.

The Iraqi Government has always insisted that its programme was peaceful in intention, and was designed to provide a source of power and export earnings for when Iraq's oil runs out. There are several reasons for doubting this, or at any rate for worrying that the programme might change direction. One is that, despite the scale of the research and training programme, no plan has been set in motion to build even one nuclear power plant. It is difficult to find a purpose for the natural uranium that Iraq purchased in the stated aims of its programme. The evident ambition of President Saddam Hussein to achieve a dominant position in the Arab world is another cause for concern. So is the fact that Iraq gives sanctuary to international, Palestine-led terrorist organizations, which makes the presence of 93 per cent enriched uranium which can be fashioned into an atomic bomb particularly worrying. In fact, the French insisted that each used-fuel assembly must be returned to France before the next supply of fuel is sent, so that Iraq would at no time have a large quantity of bomb-grade uranium.

But it turned out that the Israelis, unlike others, were not worried about the highly enriched uranium, or at any rate, not only about that. They did not even bomb the warehouse where it was stored. They bombed Osirak, the reactor. And they talked afterwards about the plutonium threat. Now, Osirak could be used to produce plutonium, as any reactor can. It could do this best if it was surrounded by a blanket of natural uranium, and this could be a reason for Iraq's purchase of uranium. However, Iraq could not do this without the IAEA inspectors, who had placed seals on Osirak and were visiting it regularly, knowing that it was being done. Of course, the Iraqi Government – present or future – could also withdraw from the NPT and bar the entry of IAEA inspectors. This would send out an alarm signal.

Sending out such a signal in these circumstances is the purpose of the IAEA inspection system. This is the best it can do. It is like a burglar alarm system: it does not prevent a burglary, but it tells everyone that a burglary is taking place. In the case of the IAEA system, it also gives the alarm well in advance, since several months at least would be required to extract plutonium and build a bomb. To take drastic action when the alarm has not sounded is to devalue the alarm system. Sigvard Eklund, the IAEA Director-General, recognized this when he said immediately after the raid: 'The Agency's safeguards system has also been attacked.' The raid weakened the authority of the IAEA just as vigilante action weakens the authority of the police.

It also blurred the important distinction between equipment and materials that can be used directly to produce an atomic bomb and those which cannot, and devalued efforts to deny those in the former category to a country like Iraq. For the message of the raid is not that Mr Begin did not want Iraq to have certain specially sensitive materials or technology. It was that he did not want it to have any nuclear reactor at all, except for a small-scale one, presumably on the grounds that it could be used to make plutonium for bombs, and that he would not permit it to have one. Presumably, this applies to other

Arab countries (though not to Egypt, evidently; Egypt has plans to acquire a nuclear reactor, and these are causing no concern in Israel). It is the policy that was practised in the Middle East long ago by King Herod, who killed babies at birth lest one of them grow up to challenge his position.

The Israeli action also drew attention, or at least should have drawn attention, to another problem, that does not concern nuclear weapons but nuclear safety. Mr Begin said after the raid that the reactor had to be destroyed immediately because it would start operating soon, and once it was operating, its destruction would spread radioactive materials across Bagdad. This is not denied. It brings home the point that reactors, even research reactors, can present a hazard in wartime. In fact, research reactors may be more of a hazard than the much larger nuclear power plants, because a power plant has a concrete shield several feet thick which would be impervious to most high explosive bombs. This was a point made by the Austrian nuclear physicist Egon Broda, when he presented a paper at a conference in 1978 called 'Destruction of Nuclear Power Plants in Crisis and War'. Broda pointed out that a nuclear power reactor produces more radioactive fission products than an atomic bomb, particularly long-lived products such as strontium 90 and plutonium.

* * *

The problem of nuclear weapons proliferation is primarily political because a decision to acquire nuclear weapons is a political act. The principal reasons that more governments have not taken this decision, and that nuclear weapons have not spread as rapidly as most people thought they would when they were first developed (see Chapter 4) is probably that international relations were fairly stable for the twenty-five years or so after 1950. Things happened during those years, and some of them seemed important at the time. But there were no really momentous changes compared with others that have taken place in the past century. The system of alliances that sprang up at the start of the Cold War held firm, in Europe there were no major changes of frontiers, and most nations felt the possibility of war and conquest recede as the years went on.

Why would a country want to acquire nuclear weapons? It may be in order to engage in direct competition with the superpowers

(China) or to be able to defend itself against conquest by one of them in a world war (Britain and France). A country might also want to have nuclear bombs in order to influence the actions of a superpower without competing directly. It has been argued persuasively that Pakistan wants to acquire nuclear weapons potential, not in order to further its rivalry with India, but in order to be able to influence superpower actions in South Asia.* The example of China indicates that this is a good way to do so. Immediately after China's explosion of a nuclear bomb in 1964, many Americans started saying that it was time the United States should stop treating China as a pariah and try to bring it into the community of nations.

A nation might want to acquire nuclear weapons because it is threatened with conquest, particularly if the conquest that is envisaged means total elimination as a political entity. Four countries come to mind that are in that position or that might be soon: Israel, South Africa, Taiwan, and South Korea. The last two of these are protected by a defence alliance with the United States, but they may well feel less than certain about the American will to come to their aid if doing so means risking a world war.

Israel, as we have seen, has the capability to manufacture nuclear weapons. South Africa will have soon: it has its own uranium supply, and will have an enrichment plant. Taiwan, an NPT signatory, has a sophisticated technology and a broad industrial base, and has an ambitious nuclear power programme. It plans to have six US-built power reactors operating by 1985. All its nuclear facilities are under IAEA safeguards. American officials were alarmed when they found that Taiwan had built a small 'hot cell' to handle plutonium in the early 1970s, and persuaded it not to go further in this direction.

South Korea also has an industrial and technological base, and a nuclear power programme, thought less advanced than Taiwan's. Its attempt to purchase a reprocessing plant from France in the early 1970s, well in advance of any industrial need, indicates – though it does not prove – that the Government was thinking of a nuclear weapons option. South Korean officials have discussed the

*The Nuclear Spread: a Third World View, by Ashok Kapur. Article in *Third World Quarterly*.

possibility with visitors from Western countries, who have usually tried to discourage them.

A country might want to develop nuclear weapons to enhance its status. India exploded its nuclear bomb for purposes of prestige, both domestically and internationally. If Pakistan explodes a weapon, it will be to demonstrate its capability. This kind of prestige is not dissociated entirely from military power. Its basis is the demonstrated ability to construct nuclear weapons if the need arose, and it implies the ability and will to use them.

In South America, Argentina and Brazil are traditional rivals for continental leadership. However, the continent has so far not had enough political homogeneity to be led, the two countries are in a period of co-operation, and in any case, Brazil is incontestably the more powerful of the two today. If either country made a move to acquire nuclear weapons, it would be in order to establish a claim to primacy. Neither faces a direct military threat. Neither needs nuclear weapons for defence or deterrence. On the other hand, a major disincentive for either to go nuclear is that the other would almost certainly try to follow suit for the same reasons of prestige.

The purpose of acquiring a weapon will decide what one does with it. If the purpose is primarily prestige, then a demonstration explosion is usually necessary. The only point of India's constructing a nuclear device was to explode it so that the world could see. If the purpose is deterrence, then a demonstration might be a good idea. If South Korea possessed a nuclear bomb, it might stage a test to deter any attack by North Korea.

For another situation, the requirements are different. If Israel has nuclear weapons, it is wiser for her not to disclose the fact. If she did, neighbouring countries would have to react. No matter how strongly they may believe now that she has nuclear weapons, they are not under the same pressure to take counter-measures as they would be if Israel made a public announcement of the fact, measures such as intensified efforts to acquire nuclear weapons of their own, and a plea for military help and perhaps nuclear protection from the Soviet Union. A senior Egyptian military figure was discussing the question recently, and talked about the dangerous consequences that would flow from a public test of an Israeli nuclear bomb. What, he was asked, if Israel has a bomb now and Egypt does not know it? 'If we don't know about it, then it wouldn't make any difference,' he replied.

Israel has no need of nuclear weapons because she is stronger than any possible combination of Arab countries that might oppose her. She has demonstrated this in successive wars. President Sadat recognized the fact when he signed the armistice agreement with Israel. The destruction of Israel may be a dream of many Arabs, but it is not on any Arab country's agenda for the forseeable future. Hence the Egyptian general's reply about the Israeli atomic bomb. Nevertheless, given the psychological pressure on Israelis created by the insistence of so many Arab figures that their country must be destroyed, and the sheer numbers opposing them, as well as the Holocaust that lies in the background of Israel and all Israelis, it would be understandable if they wanted to acquire the added insurance of a nuclear deterrent.

If Israel had nuclear weapons, a rational strategy for employing them might be to keep them in reserve, and in secret, for one awful eventuality, if it ever came. If conquest seemed inevitable, with enemy tanks rolling into the suburbs of Tel Aviv and the centre of Jerusalem, then Israel would announce that unless enemy forces withdrew, nuclear bombs would be dropped on the capitals of the countries involved. (To threaten death and suffering to civilians on this scale may seem appalling to contemplate, and a hideous perversion of the original ideals of Zionism. Nevertheless, this is precisely the threat that the Western democracies, as well as the Soviet Union, are making now with their deterrence posture.)

However, if Israel finds that it faces further wars, and possesses nuclear weapons, then there could be a shift to a different strategy. Israel is a tightly-knit country, in which most people of the same age group are linked to one another in a network of organizations. The death of a young man in battle sends tremors along this network, and is felt deeply and widely. To say this is not to suggest for a moment that such a death in any other country is felt as less of a tragedy by family and friends; only that the deaths of young Israelis may produce more political consequences.

If blood-letting continued in one war after another at intervals of a few years, such as marked the first twenty-five years of Israel's existence, it would be tempting to cut short the torment at one point by drawing a line on a map and saying that if any enemy crosses that line, he will be met with tactical nuclear weapons. Or else simply keep tactical nuclear weapons in reserve and use them if another war broke out. This would raise the level of violence and open the way to nuclear escalation. Several theories have been put forward about the mysterious flash in the South Atlantic in 1979 observed by a reconnaissance satellite which had some characteristics of a nuclear explosion but not others. It was reported that a passage in an Israeli book deleted by the military censor said this was a test of an Israeli neutron bomb, carried out with South African co-operation. This has some political plausibility. A neutron bomb is always a low-yield weapon for tactical use; if this report were true, it would mean that Israel is already creating an arsenal of nuclear weapons to be used on the battlefield, and not just for strategic deterrence.

There are political disincentives to building nuclear weapons. One is the disapproval of friends and allies. The exemplary case of political disincentive is West Germany. It is in the front line of the Cold War in Europe, and has the industrial capability to build nuclear weapons. However, if it did so, it would bring on its expulsion from NATO, and also provoke Russia. Thanks to NATO, Germany is safer without nuclear weapons than with them. Other countries also are protected by defensive alliances with the United States. South Korea had to back down from buying a reprocessing plant when the United States hinted that this would jeopardize their defensive alliance.

A country with a considerable nuclear power programme would be putting it at risk if it used its facilities to construct nuclear weapons, because this would almost certainly mean a cut-off of supplies from other countries. A leading Japanese commentator on nuclear energy, Imai Ryukichi, pointed out in an article: 'The moment Japan touches nuclear armaments, the import of materials and technology will be cut off. Our nuclear power industry, which is supposed to supply one-third of our energy requirements by 1995, will suddenly come to a halt.' If a country becomes heavily dependent on nuclear power, then it becomes dependent also on the goodwill of supplier countries, or at least the absence of ill-will, and more vulnerable to institutional penalties such as that posed by the NPT. This is why some officials concerned with proliferation are not displeased when a country with nuclear power such as Brazil decides to expand its programme and rely on it for more of its electricity: they believe that whatever increase in technical capability this involves, it makes it less likely that the country will build atomic bombs.

We live now with the danger of nuclear war. If more nations have nuclear weapons, then the number of possible nuclear wars will increase. Some people have suggested that once a nation has reached the state of scientific and technological development needed to produce nuclear weapons, then it will have attained a level of safely rational behaviour, that an Idi Amin with an atomic bomb is a nightmare that could not come true. This idea might have afforded some comfort if the years between 1933 and 1945 had not shown what can happen to a scientifically and industrially advanced country.

In fact, a number of Third World countries might be more prone to use nuclear weapons than the present possessors. For one thing, they have a lower threshold of conflict. It is hard to imagine any of the major powers going to war with another over the kind of territorial claim that led to the Iraq-Iran War. For another, in those parts of the world where nations are newer, and territorial boundaries less firmly fixed by the weight of centuries, there are more potential causes for conflict in ethnic discontent, where a group of people decides it does not belong in the nation to which the decolonizing process assigned it. For another, there has grown up among the five nuclear powers a nuclear taboo: a tacit understanding that the threshold between conventional weapons and any kind of nuclear weapon, of whatever explosive power, is to be crossed only *in extremis*, because this opens the way to nuclear escalation with no further barrier in sight. A new nuclear power might not draw the same distinction between nuclear weapons and other weapons.

In discussion of the possible spread of nuclear weapons, some people draw consolation from the fact that a new nuclear weapons state would almost certainly produce fission bombs, with a low yield of 'only' 10 or 20 kilotons, not the multi-megaton hydrogen bombs, which require a much greater scientific/technological effort to build, and which in their thousands make up the nuclear arsenals of the superpowers. This only shows how the imagination, not to say the conscience, has been dulled by the war talk of the nuclear age.

A war fought with low-yield nuclear weapons can be contemplated with something like equanimity only if it is compared with the superpower nuclear exchanges that exist in the scenarios of strategic analysts and the war plans of military staffs, rather than with anything that has actually happened. If one compares it with wars that have occurred in the past, then a conflict involving 'low-yield' fission weapons appears destructive on a terrible scale. As Albert Wohlstetter has pointed out, the atomic bombs that were dropped on Hiroshima and Nagasaki were both low yield by today's standards, less than 20 kilotons, but they terminated a rather large war.

* * *

During the last decade, governments have been considering the possibility of yet another kind of nuclear proliferation: nuclear terrorism, the acquisition of nuclear weapons by what the professional literature calls 'sub-national groups'. Concern about this is now a factor in shaping government policies: on the spread of plutonium, for instance, and on nuclear facilities in the Middle East. Though some scientists still make light of this danger, governments cannot fail to pay attention to some of the sombre warnings they have received. For instance, Lord Flowers, the physicist and Chairman of the Royal Commission that produced the report on nuclear power, said that if plutonium is used widely as a reactor fuel, then the question is 'not whether somebody will acquire it for purposes of terrorism or blackmail, but when'.

The first person to sound a strong warning about nuclear terrorism was Theodore Taylor, a distinguished theoretical physicist who designed nuclear weapons at the Los Alamos laboratory. He saw the possibilities, and he understood why others engaged in the design and fabrication of nuclear weapons did not see it. They build highly sophisticated weapons that are very efficient, with a predictable yield, small enough and strong enough to be deliverable in a missile warhead. They asked themselves whether any group of terrorists anywhere would have the resources they were employing, or even the skills, and immediately answered in the negative. But in considering the possibility of a terrorist bomb, that is the wrong question to ask. The right one is: what is the minimum that is required to produce a nuclear explosive device that would probably work even though it may not be maximally efficient or predictable in its yield, that might be too cumbersome to carry in anything smaller than a truck, if those making it were willing to cut corners on safety measures and take a few chances? The answer to this is a very different one.

Taylor wanted more cognizance taken of the possibility, and more precautions taken against it. He lobbied privately with US Government departments and with the IAEA in Vienna. He discussed the matter with a few scientific friends, and agonized over whether to go public with his fears, trying to balance in his mind the benefit of arousing public concern against the danger that this might plant the idea in the minds of terrorists who would not otherwise have thought of it. As it happens, when he decided to go public, he reached a very wide audience, for he voiced his fears in interviews for a series of articles that John McPhee wrote for the *New Yorker* as a profile of Taylor. These were published as a book under the title *The Curve of Binding Energy*. This was widely read and aroused comment and concern.

Meanwhile, Taylor got together with Mason Willrich, a lawyer and scholar who has written on several aspects of nuclear power, and did a study for the Ford Foundation of the possibility that people could steal nuclear materials and make a bomb. The result was a book called *Nuclear Theft: Risks and Safeguards*. Their conclusion on the risks is chilling:

> Under conceivable circumstances, a few persons, possibly even one person working alone, who possessed about ten kilograms of plutonium oxide and a substantial amount of chemical high explosive could, within several weeks, design and build a crude fission bomb. By a 'crude fission bomb', we mean one that would have an excellent chance of exploding, and would probably explode with the power of at least 100 tons of chemical high explosive. This could be done by using materials and equipment that could be purchased at a hardware store and from commercial suppliers of scientific equipment for student laboratories.

Public concern was given a new twist when a graduate student at Princeton, John Phillips, designed an atomic bomb for a term paper on nuclear proliferation, using material information publicly available from official sources such as the Los Alamos laboratory.

Most of the many newspaper and magazine articles about Phillips' achievement exaggerate its significance; after all, a competent graduate student in aeronautical engineering might be able to design a jet airliner, but he could not build one. Nonetheless, Phillips' supervisor, a physicist with some personal knowledge of nuclear weapons, was astounded at the amount of detailed information available, and urged Phillips to destroy his list of sources.

The critical mass of fissile material – the amount at which an explosive chain reaction begins – varies according to its purity, but it may be as little as 10 pounds for uranium that is almost 100 per cent U-235, and is usually the size of a grapefruit. Creating the fissile material is by far the most difficult part of building an explosive device. No terrorist organization could enrich uranium or create plutonium. However, one might steal one of these substances, particularly if more and more comes into use.

Highly enriched uranium is used in research reactors and HTGRS, often enriched to 93 per cent. Fifty pounds of this probably constitutes a critical mass. It is easier to make a bomb with highly enriched uranium than with plutonium. It can be achieved with two sub-critical pieces shaped so that they fit together precisely, and high explosives, and a container. Other items, such as a neutron reflector, would enhance the power of the explosion, but are not essential. Plutonium collects in one form or another at a reprocessing plant and at breeder reactors. It emerges from a reprocessing plant as a liquid, but this can be converted into the oxide powder and then into metal by a fairly simple process, though special handling equipment would be needed because of the toxicity of plutonium. Plutonium can be extracted from the fuel rods of a breeder reactor, which consists of plutonium oxide and uranium oxide, again by a chemical operation that is not beyond the competence or resources of a few skilled people with the money to buy equipment. The plutonium bomb is rather more complicated than the 'gun barrel' method that is usually used to bring two pieces of U-235 to critical mass, and must be measured more precisely, but we have the conclusion of Taylor and Willrich, and others as well, to remind us of the possibilities.

The theft of one of these materials might easily be undetected, because the amounts that result from various industrial processes are not always measurable. This is seen in the disagreement over whether highly enriched uranium was stolen from the Nuclear Materials and Equipment Corporation or whether it was lost, buried or simply not created in the first place. (The corporation was fined $1.1 million for not having a conclusive answer.) It is seen also in the disappearance of two plutonium fuel rods from the experimental fast breeder at Dounreay, which were almost certainly disposed of by mistake. In a reprocessing plant, the normal limit of error in accounting for materials is 1 per cent, which means that a small fraction of the throughput of plutonium could be stolen without anyone knowing.

The warnings have had an effect. The IAEA has laid down standards for protection of materials against theft, and requirements for anti-theft measures are sometimes written into export agreements. Sensitive fissile material is transported under military-type guard and in secret, and the precise location of stores is also kept secret.

Government agencies spends a lot of time now considering the possibilities of nuclear terrorism. Almost certainly, they devote more time to it than any terrorist organization does. An intensive study of the literature of Palestinian organizations that support terrorism has not turned up a single suggestion that nuclear weapons might be employed.* But it is better that the minds of guardians are over-active than that they lag behind the dangers.

A theft of fissile material could be an inside job, planned long in advance with the help of someone placed in a strategic position. Or else someone might be suborned: there are plenty of examples of this from the world of criminal theft. After all, the organizations that handle fissile material are large and growing more numerous, and it is unlikely that every typist and truck driver will be imbued with the same sense of seriousness and responsibility. Once the fissile material is located, even if it is well protected, then a group of attackers with the skill, dedication and bravery that marked, say, a great many wartime commando operations, would have a good chance of getting away with it.

* This is recounted in Terror on a Grand Scale by Roberta Wohlstetter, *Survival*, May/June 1976.

Many scenarios can be written for terrorists' use of some fissile material they have stolen. They might simply pretend to have manufactured a nuclear bomb, by sending the authorities the design along with proof that they have the fissile material, and use that to blackmail a government. If they make a bomb, they might deliver it to the target area by van or by motor boat. It would constitute a threat no government would be likely to defy; the explosion of a nuclear bomb of even one kiloton in the centre of a city would have a terrible effect.

Most scenarios assume that terrorists would use a nuclear bomb for blackmail purposes. This is being optimistic. Another possibility is that a group of terrorists bent on striking a blow at a particular society might not threaten to detonate it, but do so. This is the worst possibility. After all, if someone demands, 'Your money or your life', you at least have a choice.

Chapter 14

NEW ANSWERS, OLD QUESTIONS

In everything we undertake, either on earth or in the sky, we have a choice of two styles, which I shall call the grey and the green. The distinction between grey and green is not sharp. Only at the extremes of the spectrum can we say this is green and that is grey. The difference between green and grey is better explained by examples than by definition. Factories are grey, gardens are green. Physics is grey, biology is green. Plutonium is grey, horse manure is green. Bureaucracy is grey, pioneer communities are green. Self-reproducing machines are grey, trees and children are green. Human technology is grey, God's technology is green. Clones are grey, clades are green. Army field manuals are grey, poems are green.

Why should we not say simply, grey is bad, green is good, and find a quick path to salvation by embracing green technology and banishing everything grey? Because to answer the world's material needs, technology has to be not only beautiful but cheap. We delude ourselves if we think that the ideology of 'Green is Beautiful' will save us from the necessity of making difficult choices in the future, any more than other ideologies have saved us from difficult choices in the past.

Freeman Dyson, *Disturbing the Universe*

The profound changes that have come about in the management of nuclear power during the past three decades have stemmed from changes in policies, attitudes, economics and expectations, not in science or technology. Nothing new about uranium fission has been discovered, and no new technology concerning its application has been invented. Now, however, there are new developments in technology which, if they are successful, will have important consequences.

One is laser enrichment: enriching uranium by projecting laser beams of a wave length that will ionize the U-235 atoms but not the u-238 atoms, so that the two can then be separated magnetically. This has been under development for several years. Since laser technology is widespread, many people feared that the development of laser enrichment would bring u-235 bombs within the reach of many countries and even many laboratories. Exxon Nuclear Inc has been financing extensive studies of a laser enrichment process, and its heads were concerned about this danger. So they appointed a committee of distinguished scholars headed by T. Keith Glennan, a former AEC Chairman, to report on the proliferation risks.

Their report was reassuring. It said this laser separation method, 'far from being a simple technology capable of being mastered by many countries and even sub-national groups, is extraordinarily complex and difficult. Its practical application remains at least a decade away.' It went on to say that any country with the scientific and technical resources to master laser technology 'could produce weapons-usable material by several other means more easily and with greater certainty'.

Exxon officials said that the basic process has been worked out, but problems in several different technologies remain to be solved. Work is going ahead, and the US Department of Energy plans to build a pilot plant. The benefits can be considerable. Laser enrichment would be cheaper than any other kind, and more thorough; the precise atom-thin beam would separate more u-235 than any other kind of enrichment process, and could even separate some from the thousands of tons of depleted uranium which have already been through enrichment. This would increase the supply of nuclear fuel, and reduce the requirement for fast breeder reactors.

France is developing a new enrichment process that is designed specifically as an anti-proliferation measure, and the US Department of Energy is co-operating in the work. It is a method of separating uranium isotopes by chemical means alone. Its beauty, if it works, is that it can enrich uranium only so far and no farther, up to about 5 per cent, which is quite adequate for light water reactor fuel. It could never be used to make nuclear weapons material. Then, if a country says it wants to buy an enrichment plant in order to achieve nuclear independence (as Brazil has said), it could have a chemical enrichment plant without arousing any fear that this would enable it to build nuclear bombs.

France's Commissariat à l'Energie Atomique is also behind another development designed to reduce proliferation risks. This is a fuel element which uses low enriched uranium, well under 20 per cent, but because of its configuration it could serve just as well in a research reactor as the 93 per cent enriched uranium which is the normal fuel now. This was offered to Iraq, and refused. It is called 'caramel' because the uranium is arranged in the moderator in small cubes, and somebody decided that they look like pieces of caramel.

The most far-reaching development is controlled nuclear fusion, which may one day supplant fission power. It would have many advantages over fission power. Its fuel would be as plentiful as water in the oceans, it would dispense with uranium, it would create no plutonium, and would create less radioactive waste.

However, over the years, fusion power has taken on something of the character of a mirage, that seems almost within reach but that, once you grasp at it, turns out to lack substance. It also has, like many mirages, a rosy glow, through which no flaws are visible. The excessive optimism of many early pictures of nuclear power, as the energy source that would supply all the world's power needs with little cost and no problems, has been transferred, in many people's minds, to fusion energy, which is sometimes depicted in these terms. At the first Atoms-For-Peace Conference in 1955, Homi Babha said in his presidential address: 'I venture to predict that a method will be found for liberating fusion energy in a controlled manner within the next two decades.' Two years later the authorities at Harwell said they were on the verge of cracking the fusion problem. Others have held out the same prospect. Ever since Babha's forecast, it has remained twenty years away.

Fusion power is the power of the sun, the power that keeps the stars burning, the power of the hydrogen bomb. In atomic fission, a big atom splits in two so that energy is released. In atomic fusion, two hydrogen atoms, which are very small atoms, join together so that energy is released. Atoms fuse only at temperatures of millions of degrees. In a fusion bomb, or hydrogen bomb, this heat is achieved by the explosion of a fission bomb, which sets off the fusion reaction. The problem of creating a controlled fusion reaction is to find some other way to produce enough heat.

This is one area in which the Russians are probably ahead of any Western country in their work, and one of the few areas in which, for some reason, their scientists have been willing to speak openly. Back in 1955, when Khrushchev and Bulganin visited Britain, they were accompanied by Igor Kurchatov, the physicist who had headed the Soviet atomic bomb programme. When they visited the nuclear research centre at Harwell, Kurchatov addressed the British scientists, and told them all about the exciting work that was being done on fusion at his laboratory. The British fusion project was contained in a building next to the lecture hall in which he was speaking, but the work was classified. The British scientists were embarrassed at not being able to respond to his openness by telling him what they were doing.

Work on fusion power involves immensely powerful and very expensive machines. There are two principal approaches. In one, a plasma – that is, a hot, ionized gas – is created in a tube, and is contained within a magnetic force field so that it does not touch the sides and lose some of its heat. A powerful electric pulse is sent through it to raise the temperature further. The tube in which the plasma is contained is ring-shaped, and is given a Russian name, since the Russians pioneered work in this field, a tokamak, which is the Russian word for doughnut.

The other method is to direct a laser beam or a high-energy beam of protons at a tiny pellet of matter containing hydrogen, and compress it and heat it. The required temperature would be achieved only for something like a billionth of a second. The power generated for this brief period is intense.

The use of charged particle beams has strong military implications. The beams that are being developed for fusion research might also one day be used in space against satellites, or for anti-missile defence. Charged particle beams are a big field of military research and development by the superpowers. This is why Britain and France vetoed a proposal by the European Commission to set up a European charged particle beam fusion project. There is a European fusion project at Culham, in England, but it is working only on the tokamak method. As it is, some scientists working on the project believe that when a site for the project was being considered, the Soviet Government intervened to ask privately that the project not be situated in Germany.

It seems likely that hydrogen atoms (actually deuterium and tritium, two heavier isotopes of hydrogen) can be made to fuse in a tokamak for the required length of time fairly soon. The next step, and a big one, is to create a fusion process that produces more energy than is required to set it off. In a fusion power plant, this would be used to turn water into steam which would drive an electrical turbine. The US Department of Energy says a fusion power plant might be operating by the year 2025, but this can only be a guess, since the construction of one will mean solving engineering problems which have not yet even appeared, and creating materials to withstand new extremes of heat, cold, stress and radiation.

Such a plant would be without many of the features that cause anxiety in fission power plants. It would not produce any weapons-usable material. It would produce only a small amount of used fuel as radioactive waste: helium gas, which decays in months. The machinery itself would become intensely radioactive, so that it would have to be enclosed in concrete and operated by remote control, but it might be possible to use materials that would ensure that the radioactivity would not be long-lived.

The big near-term question over the development of fission power is still the fast breeder reactor, and whether the world will push ahead into a plutonium economy. The choice is not as stark as one simply between the LWR and the once-through fuel cycle or the reprocessing-plutonium-fast breeder cycle. There are other alternatives.

Other reactors use uranium more efficiently and have some of the advantages of the fast breeder in fuel saving, though not all. The CANDU uses considerably less uranium per unit of power produced than the LWR (though it also produces more plutonium) and so does the HTGR, though this latter has the disadvantage, from the proliferation point of view, that it uses highly enriched uranium. These achieve greater efficiency because within them, uranium is converted to plutonium more quickly, and some energy comes from fissioning plutonium atoms. Because of this, they are sometimes called 'converter' reactors, and are seen as half-way to a fast breeder. Research is going ahead, and there may be more, to achieve a higher conversion rate without using plutonium as fuel. Meanwhile, Japan is developing what it calls an advanced thermal reactor which uses plutonium as fuel, but very efficiently; a 250-MW prototype is already working, called Fugen – Fugen is a Buddhist deity who can control powerful things like elephants. The Japanese are reprocessing fuel at their Tokai Mura plant and feeding the plutonium into Fugen.

A more radical departure would be the adoption of the uranium-thorium cycle. This has been in view since the beginnings of atomic energy, but no one has ever actually used thorium to produce nuclear power, even experimentally. Thorium is a fertile material like U-238; that is, if subjected to a flow of neutrons, it turns into a fissile material, in this case uranium 233, an isotope of uranium that is even rarer than U-235. The thorium would have to be fed into a reactor that is already operating, so that it would be exposed to the neutron bombardment, and then some of it would turn into U-233. This, once it was separated from the rest of the used fuel in a reprocessing plant, could be fed back into the reactor. This could be a closed cycle similar to the plutonium-fast breeder one. Though U-233 is bomb material, it might be possible to 'spike' it permanently with U-238 to reduce the possibility that it could be diverted to make weapons, so that this cycle could have some proliferation advantages over the plutonium fast breeder. It would require first the construction of a plant to reprocess used thorium fuel.

The heavy water reactor might become more popular, particularly if its conversion ratio can be improved, or else some other reactor with a high conversion ratio. There is no reason to assume that the LWR's domination of nuclear power is permanent.

* * *

Looking back over the development of nuclear technology, one sees that in almost every country and in every period, there has been similar underestimation of the difficulties and the cost. (This itself is common to many new technologies which are entered into with enthusiasm, such a missile guidance and supersonic air transport.)* Problems such as the disposal of radioactive waste, and plutonium, were largely ignored for years. The impression given is that the prospect of nuclear power for the world presented such an attractive picture that features that might spoil it were brushed

* Including synthetic fuels, which have been advocated by some as one of the alternatives to nuclear power. A Rand Corporation study of forty US corporations which have built synthetic fuel plants shows that the *average* cost over-run was 300 per cent.

aside irritably, as minor details that could be attended to later, without much thought being given as to how minor they really were.

The official forecasters used sophisticated methodology, and they were wrong. It appears in retrospect – and because so many forecasters were wrong in the same way, one can generalize about this – that they were wrong not simply because they made mistakes, but because they said they knew things that they could not know, such as what engineering problems would arise and how quickly they could be solved, the evolution of uranium exploration, and the future economics of other fuels, or else they said that the things they did not know were not important. In fact, with so many uncertainties, it may well be that the only correct answers to questions about the pace and cost of nuclear power development would have been, 'I don't know', or else an estimate within such wide margins of error, and containing so many variables, as to be of little value as a guide to planning. But government departments and industrial corporations want answers to questions, and usually will not pay for an uncertain shrug and an analysis of the uncertainty.

One common mistake in recent years has been the overestimation of demand for electricity, which has meant miscalculating the economics of nuclear power. During the late 1960s, the mistake was simply extrapolating the consistent annual increase in demand that went with consistent annual economic growth; this was stopped by the oil price rise in 1973–4, and the subsequent worldwide recession. In some countries, particularly Britain, natural gas became an alternative to electricity in many uses. The overestimation has continued since then. It is natural for political leaders to forecast greater prosperity in the future; they must do so, to satisfy their citizens and, also, presumably, to keep their own spirits up. But this can distort economic planning. Britain, for instance, although it is making ambitious plans for an expanded nuclear power programme, has at the moment an excess of electrical generating capacity. In February 1981, a House of Commons Select Committee questioned the official forecasts of demand, and the plans for expansion of nuclear power.

Today, no aspect of nuclear power is likely to be neglected, if only because there are critical eyes looking at it from every angle. But this has so far produced more questions than answers, and the same questions: is it healthy to become more dependent on nuclear energy? Will we need breeder reactors before the end of the century? Will we have enough uranium without them? Will the danger of plutonium spread be greater with or without breeders? Will we be able to dispose of radioactive waste in an acceptable fashion without reprocessing? Will restrictions on nuclear trade reduce the dangers of proliferation, or will they push nations into going their own, possibly dangerous ways? Will there be other sources of power available? Will we be able to continue using more power? And so on.

The lack of any consensual answer to these questions was demonstrated strikingly in the CONAES report. The US National Academy of Sciences and the Nuclear Regulatory Commission set up a Committee on Nuclear and Alternative Energy Systems (CONAES) to examine the whole subject. The committee came to some firm conclusions, for instance, on conservation and on coal. But the points on which the committee members could not reach conclusions are the most interesting.

On nuclear power in general, the report defines two positions: one is that it should be a power source of last resort, and the other that it is the most desirable of available sources and should be expanded. The report says: 'The Committee seemed to be nearly evenly split on these positions, with some spectrum of intermediate positions.'

On fast breeders, the report defines two positions: one recommending a breeder as soon as possible, and the other recommending an indefinite delay. It says: 'The members of CONAES were about equally divided between these two positions'.

On proliferation, the two positions are that reprocessing and breeders will contribute to the spread of nuclear weapons; and that they are irrelevant and US attempts to prevent their construction will do more harm than good. Once again: 'The members of CONAES are almost equally split on this issue.' On this, the report goes on, as if by way of emphasis: 'The only recommendation which could be agreed upon was that the outcome of present policies is uncertain.'

Small wonder that Edmund Muskie said in frustration once that he was looking for a one-armed scientist, because every time he asked about nuclear power, he was told that 'On the one hand this, and the other hand that.' (Actually, there are many scientists willing to come down firmly on one side. However, they come down on different sides.)

It may well turn out that we need more nuclear power. People will not willingly see their material standard of living reduced. If a shortfall in power production means that, then it will mean a world in which nations struggle with one another ever more savagely for their share of diminishing resources. The Third World, which already has huge areas of undernourishment, will come out worst, because it is the weakest. Only intensive agriculture can feed the growing populations in the Third World, and, for that matter, the more stable (in size) populations of the rest of the world. This means the production of nitrate fertilizers, which requires power. An energy-short world will be a dangerous world, and an uncomfortable one for many.

Not knowing all the problems that lie ahead, we cannot afford to turn our backs on any aspect of technology, because it could be essential to our survival. Through the size of our populations and our patterns of living, we have become dependent on technology for life itself as much as on the natural world. Two centuries ago, most countries were largely self-sufficient in food. Today very few countries if any are self-sufficient in food and the raw materials needed to sustain life. Transport and communications are now as essential as primary products. The inhabitants of a modern metropolis are as dependent on man-made life support mechanisms as the crew of a submarine; if they were cut off from power and transport, people would start dying of starvation after a few days, and perhaps of cold also, and rats would emerge from the sewers after a week. We need technology to solve some of the problems that technology has created, such as industrial pollution.

We may need nuclear power to avoid what is called the 'greenhouse effect'. It has been discovered in the past few years that burning coal and oil may also have its dangers, in the production of carbon dioxide. The amount of carbon dioxide in the atmosphere has increased by about ten per cent since the beginning of the industrial age, and is increasing now more rapidly than ever. Furthermore, we are chopping down forests which soak up some of it. Carbon dioxide acts on the Earth much like the glass in a greenhouse. It allows the sun's rays through to warm the Earth, and traps some of the heat which would otherwise radiate out into space.

The temperature of the Earth has already increased slightly. Our present knowledge of the dynamics of the Earth's atmosphere is imprecise, and we do not know how far a further increase in the carbon dioxide content of the atmosphere will raise the temperature further, nor just what effect this would have. It is likely to have some deleterious effects: for instance, shifting the areas climatically favorable to agriculture northwards, and reducing the circulation of sea water and therefore reducing the numbers of fish and sea plants. Small changes in climate can trigger off major effects.

Scientists in this field are working hard to acquire firm data. A characteristically cautious warning is this, from a paper read to a conference on climatology held in London in 1980: 'There is no cause for alarm, but we should realize that it may not be advisable to continue the rapid increase in the use of fossil fuels. We may one day find ourselves in the position where we shall have to refrain from the full exploitation of remaining reserves.'* Nuclear power does not produce carbon dioxide, and is an obvious alternative if nations have to cut back on the use of coal and oil.

If we do need nuclear power, it may turn out that we also need the breeder reactor, for the reasons that its most zealous advocates say we do: because the uranium is running out, and because it simplifies the disposal of used fuel. Given the long lead times involved, an expanded nuclear power programme based on a network of fast breeders would only come into being twenty-five to thirty years from now, and only begin to pay back the investment forty years from now. One cannot say with certainty that a viable alternative will be found in that time, but it would be a very rash person who would say that one will not. A calculation forty years ago of power sources today would not have included nuclear fission.

There are several other possible sources of power that might be developed to the point of economic viability on a large scale within the next few years: solar power, most probably in the form of a more efficient photovoltaic cell, which converts sunlight directly into electricity; wave power; geo-thermal power; methane gas extracted from beneath the Earth's crust; hydrocarbons produced by duplicating the photosynthesis process of plants; or growing plants to produce alcohol and other fuels (no contribution to the

---

* *Man's Effect on Climate*, by Professor Bert Bolin of the University of Stockholm.

greenhouse effect from burning these, since the plants will extract as much carbon dioxide from the atmosphere as burning the resulting fuel puts into it). This is quite apart from the likely incremental improvements in the extraction and use of coal and natural gas.

A new technology of extracting minerals might even transform nuclear power and make the fast breeder unnecessary by making uranium much more available. One possibility is the development of a process that will induce some microbes to separate uranium from ore, as some microbes can already separate copper. An experimental uranium extraction plant has been built for work on this at Agnew Lake, Ontario.

Some words of Lord Flowers on the subject of alternative sources of power are worth heeding, from a lecture he gave at Temple University:

> If nuclear power is inevitable until the end of this century, it is because we have not invested in any alternatives. It should therefore be a deliberate set of policy to ensure that we do have some alternative by the year 2000. If we do then decide upon the future expansion of the nuclear option, we should do so not because it is the only option, but because it is the best option.

* * *

At IAEA headquarters in Vienna, the Economic Studies Section of the Agency has worked out a computer program with which it can advise any country on whether it will be economic for it to have nuclear power. This is the Wien Automatic Systems Planning Package, known as WASP (a tactless acronym, since the group is sometimes telling Asian and African countries that they should not have something that white Westerners have). The criteria appear to be getting more and more discouraging to potential users of nuclear power. This is despite a fundamental flaw in the system which tends to exaggerate the capacity to employ nuclear power: for the forecast of a country's economic growth, which is the most important single factor, the Government's own assessment is taken.

Using WASP, the Economic Studies Section told the Shah of Iran that his plans for nuclear power were wildly over-ambitious. His successors evidently agreed, because they pulled out of nuclear power entirely. They told Brazil that its nuclear power programme would be economic at 1974 costs; these costs have escalated since, but the Brazilian Government has not asked for a supplementary opinion. (A WASP opinion is given only when it is requested.) When the Indonesian Government produced a plan to have a nuclear power plant operating by 1986, the men in the Economic Studies Section said this would be too soon for it to be economic. They told Pakistan and Thailand also that nuclear power might be a good idea but not just yet. They told Hong Kong that if it builds a nuclear power plant now, it will probably be competitive with other power sources by the 1990s; the Hong Kong Government decided not to go ahead because of uncertainty about the colony's status after the lease from China runs out in 1997. This is all a far cry from the enthusiastic days of the early 1960s, when the IAEA was looking forward to recommending nuclear power to countries like Kenya and Cuba.

According to WASP projections, nuclear power can only make sense on a sizeable scale. Nuclear power plants have become bigger since the first ones were built, and the smallest that is now being built for commercial purposes is 600 MW. The IAEA says that no power system should rely on one power plant of any kind for more than 20 per cent of its power, because this would make it too vulnerable. This means that a nation must use 4,000 MW of electric power to justify building one nuclear power plant. Actually, the figure must be much greater, IAEA officials say, because one nuclear power plant alone would not justify the administrative and physical infrastructure that must be built up to sustain it. For instance, some countries that are considering nuclear power would have to build new harbour facilities to unload the 400-ton pressure vessel of a PWR.

Nuclear power has lost much of its glamour for the Third World, and much of its economic promise. It has become less attractive to the first and second worlds also. Expectations have not been met. Nonetheless, it is established as an intrinsic part of the power supply in many countries. Belgium and Switzerland get 25 per cent of their electricity from it. In other countries where the national contribution is much less, some regions are heavily dependent on it. Scotland, Maine and Ontario all get a quarter or more of their electricity from nuclear power.

Huge amounts of money and hardware are invested in nuclear

power, and many careers also. Its production has spawned service and ancilliary industries. Britain and France stand to earn some two billion dollars over the next ten years from reprocessing other countries' fuel. There are scientists devoting their time and talents to comparing different configuration of fuel rods in a nuclear reactor, or investigating chemical reactions in used fuel, or the effects of radiation on different kinds of geologic formations with a view to selecting an appropriate waste disposal site. There are corporations that do nothing but transport nuclear materials, and others that supply instrumentation to nuclear power plants, and others that act as brokers for nuclear materials, and others still that act as brokers for unwanted contracts for the supply of nuclear materials. The industry holds symposia and conferences, and there are periodicals and yearbooks devoted to its affairs. To change the direction of this industry, let alone reverse it if this proves desirable, would be very difficult.

People in nuclear power tend to see their industry as a sensitive organism struggling to survive in a hostile environment, exposed constantly to wounding criticism and unfair treatment by the media, doomed to suffer grievous consequences for any trivial error, whether of technology or of public relations. Opponents see it as rich, strong and ruthless, with hands on the levers of power. It is like the situation of the committed Left and Right in a witty article by Michael Frayn, in which each sees itself as a resistance movement in a country occupied by the other.

Nuclear power remains a remarkably international industry, even though the stipulations that many countries place on the use of their goods and services add red tape and cost to the already complex trade, and anxieties about its future.

Ships built for British Nuclear Fuels Ltd., specifically to carry used nuclear fuel sail halfway around the world carrying used fuel from Japanese reactors, mined in Canada and enriched in America, to be reprocessed at the BNFL plant at Windscale and the Cogema plant at La Hague, every shipment with the permission of the US and Canadian Governments. A utility – and this is an example given by the Uranium Institute in a paper – buys uranium in Australia and sends it to Canada to be converted into uranium hexafluoride, sends this to the United States to be enriched, and then sends the enriched uranium to Sweden to be fabricated into fuel rods. The utility must get the permission of every one of these governments to reprocess the used fuel, or to transfer it to another country.

In the United States, such a request must go to the State, Energy, Defense and Commerce Departments, the Arms Control and Disarmament Agency, and the Nuclear Regulatory Commission. So must any request for the purchase of sensitive nuclear material from America. This means that a routine request may wait six months for an answer. Delays of two years are not uncommon.

However, the patterns of trade are changing. The United States is losing its whip hand as the supplier of enriched uranium. By 1985, when the new Eurodif enrichment plant at Tricastin and the Urenco plant in Germany are operating, Eurodif and Urenco will be able to supply all Western Europe's enrichment needs, even without the Soviet Union. The United States is now importing uranium from Australia. If uranium becomes scarce, then America, with more nuclear power plants than any other country, will be competing with other countries for available supplies. Canada and Australia, and perhaps, one day, an independent Namibia, the world's biggest uranium exporters, will hold the key to the uses of nuclear power, along with any other countries that come to produce uranium on a large scale.

Yet the same factors create the same dilemmas. There are some tangled conflicts, along lines of dispute which criss-cross one another: between suppliers and users of nuclear fuel, over who should determine how it is used; between suppliers and users of other nuclear facilities over who should decide who may have what; between nuclear weapons states and non-nuclear weapons states over the search for a non-proliferation régime that distributes constraints equitably; between the natural desire for independence of action and controls on trade; and the conflict, if it is that, between commercial considerations and proliferation anxieties.

Sometimes, there is certainly a conflict. When Canada was trying to sell a CANDU reactor to Japan recently, the Japanese Government asked Atomic Energy of Canada Ltd., a reactor manufacturing organization, to carry out a study for it. It took six months for AECL to get the required permission from the various Canadian Government agencies even to carry out the study. Japanese officials pointed out that with this kind of restrictive atmosphere, Canada did not seem attractive as a supplier, and Japan did not buy the CANDU. A country may pay a price for giving full rein to concerns about proliferation. And there is still intense competition to sell equipment.

* * *

A leading British Government scientist once pooh-poohed the concern over plutonium by saying, 'If I were the government of a Third World country and I wanted to build an atomic bomb, using plutonium from a power reactor is the last method I would choose.'

'But,' he was asked, 'what if you were a scientist in a Third World country and the Government came to you and said it wanted you to build an atom bomb, and it *had* plutonium from a power reactor. Would you say you couldn't, or would you go ahead and do it?'

'Ah, that's different,' the scientist replied. 'I suppose I'd go ahead and try to do it. But it would be bloody difficult.'

It is true that there are other, more direct routes to the bomb than the power reactor-plutonium route. It is easier to use a reactor just to produce plutonium, pulling out the fuel rods after a short time, when the plutonium is mostly the more fissile pu-239. It is easier to use highly enriched uranium if it can be acquired. More attention should be paid to the quantities of highly enriched uranium supplied for research reactors, which is explosive material without any further treatment. But power reactors do produce plutonium, and if reprocessing spreads, the plutonium will be extracted. For many countries, it will become the one fissile material that is readily available.

Furthermore – and it is worth saying this again – obtaining fissile material is the most difficult part of building an atomic bomb. The mechanism to make it explode is simply an engineering job, one requiring a high degree of sophistication and precision, but not the effort that is needed to enrich uranium or extract plutonium. The design for this mechanism could be worked out in advance, and most of it could even be constructed. This could be done by a government that wants to keep options open, or by a group of scientists or technicians, perhaps within the military, who want to give their government a nuclear weapons option with or without its knowledge.

This is not a hypothetical matter. It is probably true that no country in the world, with the exception of Pakistan and possibly South Africa, is working today to create nuclear bombs where it did not have them before. However, it is very probable that some people in many countries are bearing in mind the possibility that their country might one day find itself in a position where it wants to acquire nuclear weapons, that they are making calculations simply in the spirit of prudence, and even that this consideration is a minor, even if unspoken, input in plans for nuclear power. When an American television company wanted to know whether John Phillips' design for an atomic bomb would work, and the Pentagon refused to comment, it took the design to the Swedish Ministry of Defence, which came up with an authorative answer.

There are those who suggest that there are now so many ways for a country to acquire an atomic bomb that it is not worth placing impediments on the nuclear power industry in order to block this one. There are also people who say that we will have to reprocess used fuel whether we like it or not because there is no safe way to dispose of it otherwise, or that we will have to build breeder reactors because the only safe thing to do with plutonium is burn it in breeders. The logic of these arguments leads to the conclusion, not that one kind of nuclear power is safe, but that all kinds are equally dangerous; not that nuclear power should be set free, but that it should be abolished, or at any rate, that it should not have been started. As an anonymous citizen said, quoted by one scientist in a discussion. 'First they told us there was no problem. Now they tell us there's no solution.'

Some barriers must be erected between nuclear power and nuclear bombs, even if these are not insurmountable. Restrictions on the trade in nuclear materials is one way to accomplish this. The conflicts inherent in these can only be resolved, or at least partly resolved, by the replacement of rules imposed by one country or a small group of countries with rules agreed to by a large number of countries.

A country does not willingly accept restrictions that are imposed upon it by others. Also, the wider the agreement on any set of rules, the more effective their enforcement becomes. Namibia and South Africa export quantities of uranium. A country might avoid the conditions on uranium supply that are laid down by Canada and Australia by buying from one of these. Other suppliers will emerge.

Eurodif and Urenco sell enrichment to Brazil, and both are expanding their capacities.

Ideally, new rules would be amendments to the Non-Proliferation Treaty, and would be accompanied by efforts to bring in more countries. But so many NPT signatories are dissatisfied with the workings of the treaty now, and with the way that the big powers have fulfilled their obligations under it, that there is no possibility of a general agreement to tighten it.

Committees within the IAEA are discussing several plans to internationalize the acquisition and disposal of nuclear fuels. One harks back to President Eisenhower's original atoms-for-peace proposal: an international nuclear fuel bank, which would not be under any one country's control, from which any country that observed the rules might draw enriched uranium. This could conceivably give nuclear power users a source of supply that could not be choked off by a change of policy by any one country, and so might reduce the pressure to turn to the fast breeder for fuel self-sufficiency. However, the schemes that are being discussed now are too limited in scope to provide any real security of supply. They envisage the major suppliers contributing only small amounts to the fuel bank, so that the bank would only have enough to serve as a stop-gap in the case of an interruption in fuel supply.

It is also suggested that agreements to supply nuclear fuel should not be commercial contracts but international treaties. Then to alter the terms unilaterally would be to break a treaty; a government would hesitate before doing this, and the treaty could not be changed by domestic legislation.

Another plan is for international plutonium storage. Under this scheme, plutonium coming out of reprocessing plants would be handed over to an international control body, which would hand some back when that amount was needed for fuel. The excess plutonium would not be under national control, and therefore would not be available for turning into weapons. International control might be established on site: that is, there could be IAEA security guards and seals at a national reprocessing plant, at Windscale, or Tokai Mura, for instance. The countries possessing reprocessing plants, which would be the ones to surrender control, are showing some willingness to go along with this.

Still another plan, more ambitious still, is for the international management of spent fuel. Under this scheme, all the used fuel would be handed over to a national authority, which would then supervise its storage or, where there was a demonstrated need for it, reprocessing. Again, it is envisaged that the control would take place on site, or at any rate at a few national centres, not that the used fuel would all be sent to one huge international centre.

The advantage of a set of rules is that it operates automatically. It is not necessary for a government to ask whether they may or may not do certain things. It can known in advance what it can do and plan accordingly. But the proliferation issue is not static. It is fluid, and will remain so. No set of rules will provide a once-and-for-all solution. There are too many possibilities of political and technical change.

Changes in resource availability might mean that countries that are now dependent on others for certain materials may come to have sources of their own. There are technical changes mentioned earlier in this chapter which might make a non-proliferation regime easier to enforce: chemical enrichment which cannot enrich to weapons grade, research reactor fuel which is not highly enriched, co-processing, which does not produce pure plutonium. Other developments in the future may make non-proliferation rules more difficult to enforce: new and simpler methods of enrichment, for instance.

There will also be political changes which may require a response. These will be less easy to delineate, and the response perhaps less easy to justify. A country might acquire what appears from the outside to be a new motivation to build a bomb, or a new will to do so. For instance, it might leave a defensive alliance, or might find that the protection afforded by the alliance is crumbling, and strike out on a new and intensely nationalistic path. A country might appear to be a bad risk from a proliferation point of view whatever economic justification it can provide for obtaining sensitive materials, and whatever international safeguards it accepts which can always be abrogated in any case.

The basis of anti-proliferation policy is political will. In considering the possibilities of international violence, the spread of nuclear weapons is the most dangerous factor, not because it increases the risk of war, but because it can magnify the consequences. Its prevention should be given over-riding importance. International safeguards cannot by themselves prevent anything from happening, they can only provide an early warning. This is only useful if action is taken. And the safeguards are only useful if a country believes that action will be taken if it breaches them.

Yet all too often, other factors take priority over anti-proliferation concerns, so that the message spread abroad is that whatever was said in advance, a country will not pay much of a penalty for moving into weapons-potential areas in nuclear power. France may have worried about selling highly enriched uranium to Iraq, but its concern over oil supplies took priority. The United States drew back for years from carrying out its commitment to non-proliferation strictures and denying India fuel for its Tarapur reactor, for reasons connected with US-Indian relations. It sent military aid to Pakistan despite Pakistan's clandestine nuclear project and despite the Symington Amendment because, as a State Department spokesman put it, 'the threat of Soviet expansionism outweighs the risk of Pakistan detonating a bomb'. The United States used its leverage over Israel to impose an armistice in the Yom Kippur War in 1973, but has not used it to pressurize Israel into accepting international safeguards over its nuclear activities. Countries that have not signed the NPT suffer no penalty for not having done so.

Anti-proliferation concerns can be an input in many foreign policy decisions. Technical and political factors are intertwined. Whether another country or several countries build atomic bombs depend on the ease with which it can be done, the economic and political cost of doing so, the benefits to be gained, and the disadvantage or danger of not doing so.

The safer the world is, the less incentive there will be for more countries to build nuclear weapons. The less incentive there is for more countries to build nuclear weapons, the safer the world will be.

Suggested Further Reading

A complete bibliography would be a long and, to my mind, unhelpful list of books, some of which connect only at one or two points with my subject matter. Instead, I am listing a few of the books that I consulted for background purposes which might be of interest to the reader who wants to pursue one or more questions further.

For a straightforward exposition of how nuclear power works, *Nuclear Power* by Walter Patterson can be recommended, even though the space it devotes to accidents reflects the author's viewpoint. A British Government publication, the Sixth Report of the Royal Commission on Environmental Pollution, *Nuclear Power and the Environment*, also contains an admirably comprehensive and lucid account.

The current scientific view of the origins of the universe is explained in *The First Three Minutes* by Stephen Weinberg, a good account written for the general reader who is prepared to make an effort. *The Runaway Universe* by Paul Davies is an excellent book that tells the same story imaginatively. Many books chronicle the discovery of nuclear fission. I drew partly on three accounts by participants: *Leo Szilard: His Version of the Facts* by Gertrude Weiss Szilard and Spencer R. Weart; *What Little I Remember* by Otto Frisch; and *Les Rivalités Atomique* by Bertrand Goldschmidt, which bears also on events in the 1950s and '60s. There is no full account of the extraordinary post-war hunt for uranium, but *The Atom Hunters* by Harald Steinert, translated from the German, has some good information on the early days, while some of the flavour of the Australian uranium rush is conveyed in *The Uranium Hunters* by Ross Annabell, one of the hunters. *Canada's Nuclear Story* by Wilfred Eggleston tells some of what happened in Canada. On the efforts of governments to retain uranium supplies, the British and American official histories contain the best accounts: *Britain and Atomic Energy* by Margaret Gowing, particularly volume two covering the years 1945–52, *Independence and Deterrence*; and *The New World: a History of the AEC* by Oscar E. Anderson and Richard G. Hewlett.

There is surprisingly little literature on the atoms-for-peace programme, though *The Peaceful Atom In Foreign Policy* by Arnold Kramish, published in 1963, gives a contemporary account of the major issues, while biographies and memoirs by and about President Eisenhower and AEC Chairman Lewis L. Strauss are source material. So are two 1960 US Government publications: *Review of International Atomic Policies and Programs of the United States: Report to the Joint Committee on Atomic Energy, US Congress*, by Robert McKinney; and *Background Material for a Review of the International Atomic Policies of the United States: a Report to the Joint Congressional Committee on Atomic Energy by the AEC*, in five volumes. On developments in America and Britain, *Nuclear Power and the Energy Crisis* by Duncan Burn is useful. *Light Water* by Claude Derian and Irvin C. Bupp is a sceptical and meticulously researched look at the over-enthusiastic spread of nuclear power during the 1960s and early '70s.

The books on nuclear proliferation problems would fill a very long shelf. Among the more recent books on the subject which contain worthwhile information and ideas are *Nuclear Arms In the Third World* by Ernest W. Lefever, a product of the Brookings Institution; and *Asia's Nuclear Future* edited by William Overholt. Others which focus on more specific situations are *Israel and Nuclear Weapons* by Fuad Jabber, which predates recent developments but contains a lot of the history; *India's Nuclear Option* by Ashok Kapur; and *The Nuclear Axis* by Zdenek Cervenka and Barbara Rogers, a tendentious but informative book which covers West Germany, South Africa and Israel.

There is one book about the uranium cartel: *Yellowcake: the International Uranium Cartel*, by June H. Taylor and Michael D. Yokell, while some of the cartel's conduct of business is told in fascinating detail in the first of two volumes of *Hearings on the International Uranium Cartel* before the Subcommittee on Oversight and Investigations of the House of Representatives Commerce Committee, published by the US Congress.

On the changes in governmental policies regarding plutonium, in addition to the studies mentioned in the text which played a part in shaping policies, a book which shows the two sides of the argument in sharp focus is *Nuclear Energy and Nuclear Proliferation: Japanese and American Views*, by Ryukichi Imai and Henry S. Rowen. *Le Complèxe Atomique* by Bertrand Goldschmidt contains a meaty account of some of these issues, with the virtues of involvement rather than detachment. The short collection of papers called *Nuclear Policies: Fuel Without the Bomb* by Albert Wohlstetter, Victor Gilinsky and others, contains much useful analysis.

In other areas, *Grounds for Concern*, edited by Mary Elliott, is a series of essays opposing uranium mining in Australia which contains a lot of information. *Windscale Fallout* by Ian Breach deals with the Windscale inquiry and its ramifications. *Nuclear Waste* by Mason Willrich gives the parameters of that issue. The proceedings of the annual symposia of the Uranium Institute, published each year in a separate volume, contain useful material on several aspects of uranium and nuclear power.

There are as many papers and monographs on these subjects as there are books. I will mention only a few on which I have drawn. Several interesting papers were published by the International Consultative Group on Nuclear Energy that was established under the auspices of the Rockefeller Foundation and the Royal Institute of International Affairs, particularly *Peaceful Nuclear Relations: A Study of the Creation and Erosion of Confidence* by Myron B. Kratzer and Bertrand Goldschmidt, an excellent account of the evolution of export policies and controls and some of the consequences; and *World Nuclear Energy Paths* by Thomas J. Connolly and others. Two other papers that deal with the plutonium problem are *Multinational Arrangements for the Nuclear Fuel Cycle* by Ian Smart, published by the British Department of Energy, which apart from its other virtues contains some valuable tables of information not easily accessible elsewhere; and *Uranium, Non-Proliferation and Energy Security* by Steven J. Warnecke, published by the Atlantic Institute for International Affairs.

Index